Lecture Notes in Mathematics

Edited by A. Dold, Heidelberg and B. Eckmann, Zürich

T0232324

363

Conference on the Numerical Solution of Differential Equations

Dundee 1973

Edited by G. A. Watson
University of Dundee, Dundee/Scotland

Springer-Verlag
Berlin · Heidelberg · New York 1974

AMS Subject Classifications (1970): 65-02, 65Lxx, 65Mxx, 65Nxx, 65P05

ISBN 3-540-06617-9 Springer-Verlag Berlin · Heidelberg · New York
ISBN 0-387-06617-9 Springer-Verlag New York · Heidelberg · Berlin

This work is subject to copyright. All rights are reserved, whether the whole or part of the material is concerned, specifically those of translation, reprinting, re-use of illustrations, broadcasting, reproduction by photocopying machine or similar means, and storage in data banks.

Under § 54 of the German Copyright Law where copies are made for other than private use, a fee is payable to the publisher, the amount of the fee to be determined by agreement with the publisher.

© by Springer-Verlag Berlin · Heidelberg 1974. Library of Congress Catalog Card Number 73-21203. Printed in Germany.

Offsetdruck: Julius Beltz, Hemsbach/Bergstr.

Foreword

For the 4 days July 3 - 6, 1973, at the University of Dundee, Scotland, over 230 people from 29 countries attended a conference on the Numerical Solution of Differential Equations. This was the 5th in a series of biennial conferences in numerical analysis, originating in St Andrews University, and held in Dundee since 1969.

It has been our continuing aim, where possible, to make some contribution towards closing the 'communications gap' existing between those workers concerned with essentially theoretical problems of numerical analysis, and those involved in the more practical 'real world' problems to which this theory may or may not be applicable. Towards this end, it was our hope that the choice of invited speakers would appeal to as large a cross-section as possible of people concerned with the numerical solution of differential equations, including those whose primary object is to obtain numbers. Invitations to present talks were accepted by 20 eminent workers, and their papers appear in these notes. In addition to the invited papers, short contributions were solicited, and 43 of these were presented at the conference in parallel sessions. In previous years, it has been the practice to include submitted papers in the conference proceedings; however the large number of talks, and the inevitable lengthy refereeing process which would be necessary, led to the decision not to include any of these here. A separate list of the submitted papers is given, and I will be happy to supply the address of any author to interested readers.

I would like to take this opportunity to thank all speakers, including the after-dinner speaker at the conference dinner, Dr J H Wilkinson, all chairmen and participants for their contributions. It is also a pleasure for me to acknowledge here the very considerable work done in organising this conference by Dr John Morris, assisted by many other members of the Department of Mathematics, both staff and students, too numerous to mention by name. John has looked after the organisation of a number of these conferences over the past few years, and has been largely responsible for their continued success.

The typing of the various documents associated with the conference, and some of the typing in this volume has been done by secretaries in the Department of Mathematics, Mrs S Addison, Miss R Dudgeon and Miss F Duncan; this work is gratefully acknowledged.

G. A. Watson

Dundee, September 1973

INVITED SPEAKERS

de Boor, C.	Mathematics Research Centre, University of Wisconsin, Madison, Wisconsin 53706, U.S.A.
Ciarlet, P.	Laboratoire Central des Ponts et Chaussees, 58 Boulevard Lefebvre, 75732 Paris, Cedex 15, France.
Collatz, L.	Institut für Angewandte Mathematik, Universitat Hamburg, 2 Hamburg 13, Rothembaumchausse 67/69, West Germany.
Daniel, J.	Mathematics Department, University of Texas, Austin, Texas 78712, U.S.A.
Gourlay, A.R.	Mathematics Department, University of Loughborough, Loughborough, Leicestershire.
Hull, T.E.	Computer Science Department, University of Toronto, Toronto 181, Canada.
Kreiss, H.O.	Computer Science Department, Uppsala University, Uppsala, Sturegatan 4B2TR, Sweden.
Lambert, J.D.	Mathematics Department, University of Dundee, Dundee, Scotland.
Mitchell, A.R.	Mathematics Department, University of Dundee, Dundee, Scotland.
Morton, K.W.	Mathematics Department, Reading University, Reading, Berkshire.
Pereyra, V.	Department de Computacion, Facultad de Ciencias, Caracas University, Apartado 59002, Caracas, Venezuela.
Stetter, H.J.	Institut für Numerische Mathematik, Technische Hochschule Wien, A1040 Wien, Karlplatz 13, Austria.
Strang, G.	Room 2-271 Mathematics Department, Massachusetts Institute of Technology, Cambridge, Massachusetts, U.S.A.
Swartz, B.	University of California, Los Alamos, Scientific Laboratory, P.O. Box 1663, Los Alamos, New Mexico, U.S.A.
Thomee, V.	Department of Mathematics, University of Gothembourg Gothembourg, Sweden.
Wachspress, E.	General Electric Company, Schenectady, New York, U.S.A.
Wendroff, B.	Mathematics Department, University of Oslo Blindern, Oslo, Norway.

Wilkinson, J.H.	National Physical Laboratory, Teddington, Middlesex.
Young, D.M., Jnr.	Mathematics Department, University of Texas, Austin, Texas 78717, U.S.A.
Zienkiewicz, O.C.	Department of Civil Engineering, University of Wales, Singleton Park, Swansea, SA2 8PP
Zlamal, M.	Technical University, Obrancu Miru 21, Brno, Czechoslavakia.

Submitted Papers

A.L. Andrew: Solution of an eigenvalue problem from Astrophysics

A. Brandt: Generalized local maximum principles for finite-difference operators

F. Brezzi: On the approximation of biharmonic problems by finite element methods of hybrid type

H. Brunner: Generalized collocation methods for stiff systems of ordinary differential equations

D.M. Cruickshank and K. Wright: Computable error bounds for ordinary differential boundary value problems using collocation methods

A.R. Curtis: High order Runge-Kutta formulae

L.M. Delves and F.A. Musa: Weakly asymptotically diagonal systems

B.L. Ehle: Some results on step size control when solving stiff equations

H. Engels: General Runge-Kutta procedures for systems of differential equations

F.F. Fambo: A recursive method for the solution of band poly-diagonal systems

W. Förster: A note on numerical schemes in conservation law form

J. Gergely: A finite numerical solution of the elliptic boundary value problems

J.A. Gregory and R.E. Barnhill: On blending function interpolation and finite element basis functions

P. Harley: An application of the collocation method to a control problem

A. Hollingsworth, E. Doron, B.J. Hoskins and A.J. Simmons: Experiments on the use of a semi-implicit time scheme in global atmospheric models

P. Holyhead: Multistep methods for solving linear Volterra integral equations of the first kind

L. Johnston and R. Mathon: A boundary method for elliptic boundary value problems

D.P. Laurie: Automatic stepsize control in the solution of parabolic equations by exponential fitting in time

A. Lerat and R. Peyret: On the origin of oscillations in the computed discontinuous solutions of non-linear hyperbolic equations

N.K. Madsen, J.S. Chang and A.C. Hindmarsh: The numerical solution of partial differential equations with a stiff ordinary differential equation integration

M. Mäkelä, O. Nevanlinna, A.H. Sipsilä: On exponentially fitted multistep methods generated by generalized Hermite-Birkhoff interpolation

G. Marshall: Semi-direct solution of the Navier-Stokes system as a coupled pair of Poisson type equations

C.A. Micchelli and W.L. Miranker: Asymptotically optimal approximation

D.D. Morrison: The practical determination of the optimal step size distribution

S. McKee and T. Joslin: A quasi-linear system of parabolic differential equations arising from mass transfer with chemical reaction in electro-chemistry

N.R. Nassif: Numerical solution of the heat-equation by Galerkin-Generalized Crank-Nicolson

W. Niethammer: A remark on a relaxed SOR-method

S.P. Norsett: Multiple Padé approximation for the exponential function

F.A. de Oliveira: Shooting methods and two-point boundary value problems

N. Papamichael: A numerical conformal transformation method for harmonic mixed boundary value problems in simply-connected domains

G. Pierini: Analysis of round-off errors in the sparse LU factorization for matrix iteration

A. Prothero and A. Robinson: On the trapezoidal and midpoint rules for solving stiff ordinary differential equations

R. Rautmann: On iterative methods for special systems of partial differential equations

H.H. Robertson and J. Williams: Prediction, error estimation and interpolation in the numerical integration of stiff differential equations

T. Robertson: On a two-point boundary value problem

J.B. Rosser: Potentials between charged plates

B.D. Sleeman and A. Källstrom: Numerical analysis of multi-parameter eigenvalue problems

J.H. Verner and S.L. Henderson: Optimizing the range of stability in a class of hybrid methods

S.J. Wade and R.A. Sack: The wave equation in ellipsoidal co-ordinates

I.A. Watson: An analogue of the A-stability result for special second order equations

J.M. Watt: New hybrid methods for ordinary differential equations

J.R. Whiteman and J.A. Gregory: Mesh refinement of finite element methods

P.E. Zadunaisky: On the estimation of errors propagated in the numerical integration of ordinary differential equations

CONTENTS

R. BARTELS and J.W. DANIEL: A Conjugate Gradient Approach to Nonlinear Elliptic Boundary Value Problems in Irregular Regions 1

C. DE BOOR: Good Approximation by Splines with Variable Knots II 12

P.G. CIARLET: Conforming and Nonconforming Finite Element Methods for Solving the Plate Problem 21

L. COLLATZ: Discretization and Chained Approximation 32

A.R. GOURLAY: Recent Developments of the Hopscotch Idea 44

T.E. HULL: The Development of Software for Solving Ordinary Differential Equations 55

H-O. KREISS: Boundary Conditions for Hyperbolic Differential Equations 64

J.D. LAMBERT: Nonlinear Methods for Stiff Systems of Ordinary Differential Equations 75

A.R. MITCHELL and R. McLEOD: Curved Elements in the Finite Element Method ... 89

K.W. MORTON: The Design of Difference Schemes for Studying Physical Instabilities 105

V. PEREYRA: Variable Order Variable Step Finite Difference Methods for Nonlinear Boundary Value Problems 118

H.J. STETTER: Cyclic Finite-Difference Methods for Ordinary Differential Equations 134

G. STRANG: The Dimension of Piecewise Polynomial Spaces, and One-Sided Approximation 144

B. SWARTZ and B. WENDROFF: The Comparative Efficiency of Certain Finite Element and Finite Difference Methods for a Hyperbolic Problem 153

V. THOMEE: Spline-Galerkin Methods for Initial-Value Problems with Constant Coefficients 164

D.M. YOUNG: On the Accelerated SSOR Method for Solving Elliptic Boundary Value Problems 176

E.L. WACHSPRESS: Algebraic-Geometry Foundations for Finite-Element Computation 177

B. WENDROFF: Spline-Galerkin Methods for Initial-Value Problems with Variable Coefficients 189

O.C. ZIENKIEWICZ: Constrained Variational Principles and Penalty Function Methods in Finite Element Analysis 207

M. ZLÁMAL: Finite Element Methods for Parabolic Equations 215

A CONJUGATE GRADIENT APPROACH TO NONLINEAR ELLIPTIC BOUNDARY VALUE PROBLEMS IN IRREGULAR REGIONS

Richard Bartels[1] and James W. Daniel

1. Introduction

The conjugate gradient method was developed for solving systems of linear equations by Lanczos, Hestenes, and Stiefel [13,12,11,16,17,18; see also 10]. Although demonstrated to be a very powerful method for solving finite difference approximations to differential equations [7], the method has not been as popular as such other methods as SOR or ADI in recent years; however, interest in the procedure for linear problems has been reawakened by some recent modifications due to Reid [14,15]. The conjugate gradient method for nonlinear problems was first suggested by Fletcher and Reeves [8] and first analyzed theoretically by Daniel [4,5]; its impact on algorithms for optimization has been enormous, and its behavior is now rather well understood. In this paper we take an idea from Daniel [4] and exploit it so as to produce efficient procedures based on a conjugate gradient method to generate approximate solutions to nonlinear elliptic boundary value problems over irregular regions. The procedures are of wide applicability and generality and appear to be competitive with some of the best methods presently in use.

In the next section we sketch the general approach, the theory behind it, and the convergence rates to be expected. In Section 3 we illustrate these ideas via a particular implementation of the general method. Section 4 discusses the computational and programming details of our implementation and presents the results of some numerical experiments.

2. The General Procedure

For conceptual clarity and technical simplicity, we first consider linear problems. To indicate the general functional analytic nature of the approach, in this section we shall use rather general operator notation rather than a specific differential operator.

We wish to solve a linear elliptic boundary value problem over a "nice" domain D; without significant loss of generality, we consider only homogeneous Dirichlet problems. Let the equation be posed as

---

[1]Dr. Bartels is Assistant Professor of Computer Sciences and Senior Research Mathematician at the Center for Numerical Analysis, The University of Texas, Austin, Texas.

This research is supported in part by the Office of Naval Research under Contract N00014-67-A-0126-0015, NR044-425; reproduction in whole or in part is permitted for any purpose of the United States government.

(2.1) $$Mu = k$$

where M is a self-adjoint elliptic differential operator of second order, defined on $C_0^\infty(D)$, the infinitely differentiable functions of compact support in D. Letting $\langle\cdot,\cdot\rangle$ denote the usual inner product in $L^2(D)$, we then have that

(2.2) $$a\langle u,Bu\rangle \leq \langle u,Mu\rangle \leq A\langle u,Bu\rangle$$

for every u in $C_0^\infty(D)$ and for some positive constants a and A, where B is the negative of the Laplacian operator. If we complete $C_0^\infty(D)$ with respect to the inner product [,], where $[u,v] = \langle u,Bv\rangle$, we arrive at the Sobolev space $W_0^{1,2}$ and a bilinear form $Q_M(u,v)$ which satisfies $a[u,u] \leq Q_M(u,u) \leq A[u,u]$ and which equals $\langle Mu,v\rangle$ on $C_0^\infty(D)$. Thus $Q_M(u,v) = [\tilde{M}u,v]$ for some bounded linear operator $\tilde{M}: W_0^{1,2} \to W_0^{1,2}$; intuitively, $\tilde{M}u = B^{-1}Mu$, and this is in fact valid for smooth enough u.

To find a weak solution to $Mu = k$, let $\tilde{k}$ be the weak solution to $Bu = k$; thus $\langle\tilde{k},Bv\rangle = \langle k,v\rangle$ for all v in $C_0^\infty(D)$. If $\tilde{u}$ satisfies $\tilde{M}\tilde{u} = \tilde{k}$, then for all v in $C_0^\infty(D)$ we have $\langle k,v\rangle = \langle\tilde{k},Bv\rangle = [\tilde{k},v] = [\tilde{M}\tilde{u},v] = [\tilde{u},\tilde{M}v] = [\tilde{u},B^{-1}Mv] = \langle\tilde{u},Mv\rangle$, which says that $\tilde{u}$ is the weak solution to $Mu = k$.

We have of course said nothing new here; we have simply pointed out that solving $Mu = k$ is equivalent to solving $\tilde{M}\tilde{u} = \tilde{k}$ where $\tilde{k}$ solves $B\tilde{k} = k$. Roughly speaking, we have merely replaced $Mu = k$ by $B^{-1}Mu = B^{-1}k$. We shall now solve the equation $\tilde{M}u = \tilde{k}$ by the following conjugate gradient method in $W_0^{1,2}$.

(2.3) Choose u_0 in $W_0^{1,2}$. Let $p_0 = r_0 = \tilde{k} - \tilde{M}u_0$. For $n \geq 0$, let $u_{n+1} = u_n + c_np_n$ where c_n is chosen so that $[\tilde{k} - \tilde{M}u_{n+1}, p_n] = 0$. Let $r_{n+1} = \tilde{k} - \tilde{M}u_{n+1}$ and let $p_{n+1} = r_{n+1} + b_np_n$ where b_n is chosen so that $[p_{n+1},\tilde{M}p_n] = 0$.

According to the results of Daniel [4], u_n converges to the desired solution u at least geometrically with a geometric convergence factor of at least $\dfrac{1-\sqrt{\frac{a}{A}}}{1+\sqrt{\frac{a}{A}}}$; more precisely,

$$[u_n-u,u_n-u]^{\frac{1}{2}} \leq \sqrt{\frac{A}{a}}\,[u_0-u,u_0-u]^{\frac{1}{2}}\,\frac{2\left(1-\frac{a}{A}\right)^n}{\left(1+\sqrt{\frac{a}{A}}\right)^{2n}+\left(1-\sqrt{\frac{a}{A}}\right)^{2n}} \leq \sqrt{\frac{A}{a}}\left[\frac{1-\sqrt{\frac{a}{A}}}{1+\sqrt{\frac{a}{A}}}\right]^n [u_0-u,u_0-u]^{\frac{1}{2}}$$

We wish to have a simpler representation of the algorithm in Equation 2.3. If all of the iterates are sufficiently smooth, or if one interprets our symbols rather intuitively, or if one thinks of B and M as discrete approximations to the true operators, then we can interpret Equation 2.3 as follows by writing $\tilde{M} = B^{-1}M$ and $[u,v] = \langle u,Bv\rangle$:

(2.4) Choose u_0. Let $R_0 = k - Mu_0$ and let $p_0 = r_0$ solve $Br_0 = R_0$. For $n \geq 0$, let $u_{n+1} = u_n + c_np_n$ where c_n is chosen so that $\langle k - Mu_{n+1}, p_n\rangle = 0$. Let $R_{n+1} = k - Mu_{n+1}$, let r_{n+1} solve $Br_{n+1} = R_{n+1}$ and let $p_{n+1} = r_{n+1} + b_np_n$

where b_n is chosen so that $\langle p_{n+1}, Mp_n \rangle = 0$

In this form we see that we can solve general linear elliptic problems by solving instead a sequence of Poisson problems $Br = R$ by any method, such as Green's functions, finite elements, finite differences, et cetera; the sequence will converge geometrically as described above. It was remarked by Daniel [4] that this gives a useful computing tool whenever one can solve the Poisson equation easily, such as when D is a sphere, an ellipsoid, et cetera. Recently, however, many very efficient methods have been found for solving the Poisson equation numerically in fairly irregular regions. Thus, our general algorithm is to solve $Mu = k$ via Equation 2.4 using some efficient Poisson solver at each step. In the next sections we shall select a particular Poisson solver and describe the operation of the hybrid algorithm that results; this is merely to give a specific example of the general procedure. Before proceeding, however, we indicate how one can extend the process to handle nonlinear problems.

Let a nonlinear elliptic problem be denoted by the equation $J(u) = 0$, where J has the symmetric Frechet derivative J'_u satisfying $a\langle v, Bw\rangle \leq \langle J'_u v, w\rangle \leq A\langle v, Bw\rangle$ for $a > 0$. As first done by Daniel [4], we can imitate the arguments above for this nonlinear problem, generating an algorithm similar to that of Equation 2.3; interpreting the method as in Equation 2.4, we arrive at the following method.

(2.5) Choose u_0. Let $R_0 = -J(u_0)$, solve $Br_0 = R_0$ and let $p_0 = r_0$. For $n \geq 0$, let $u_{n+1} = u_n + c_n p_n$ where c_n is chosen so that $\langle J(u_{n+1}), p_n\rangle = 0$. Let $R_{n+1} = -J(u_{n+1})$, let r_{n+1} solve $Br_{n+1} = R_{n+1}$, and let $p_{n+1} = r_{n+1} + b_n p_n$ where b_n is chosen appropriately.

In the above description, we left the determination of b_n undefined; various determinations, all equivalent when J is linear, give global convergence and the same asymptotic rate. In particular we can consider, for example, $b_n = -\langle r_{n+1}, J'_{u_{n+1}} p_n\rangle / \langle p_n, J'_{u_{n+1}} p_n\rangle$, or $b_n = \langle r_{n+1}, R_{n+1}\rangle / \langle r_n, R_n\rangle$, or $b_n = \langle r_{n+1}, R_{n+1} - R_n\rangle / \langle r_n, R_n\rangle$. Generally one only chooses c_n so as to approximate the solution of $\langle J(u_{n+1}), p_n\rangle = 0$; c_n predicted by one step of Newton's method starting with $c_n = 0$ maintains the same convergence rate, for example. As in the linear case, convergence is at least geometric, with ratio $\dfrac{1-\sqrt{\frac{a}{A}}}{1+\sqrt{\frac{a}{A}}}$.

3. Using a Direct Solver for the 5-point Discrete Laplacian

As we saw generally in Equations 2.4 and 2.5, we can solve linear or nonlinear elliptic problems by solving a sequence of Poisson equations. We now consider approximately solving the Poisson equation by the standard 5-point approximation, in turn solved by one of the recent direct methods [2,3,6]. To be specific we consider a simple type of mildly nonlinear problem in two dimensions, namely

$$(3.1)\qquad (su_x)_x + (tu_y)_y = f(x,y,u) \quad \text{in } D$$

where $s = s(x,y)$ and $t = t(x,y)$ are in, say, $C^1(D)$ and $\alpha \geq \frac{\partial f}{\partial u} \geq 0$ for some α for all u and all (x,y) in D; one could easily allow s and t to depend on u as well. We assume that there exist constants $A \geq a > 0$ such that $a \leq s(x,y) \leq A$ and $a \leq t(x,y) \leq A$ for all (x,y) in D. On D we impose a square mesh D_h of uniform width h, and we suppose that each point C in the mesh either lies in the boundary of D or is such that its north, south, east, and west neighbors in the mesh, denoted N, S, E, and W, are also in D or the boundary of D. As an approximate solution to $-\nabla^2 u = f$ in D with $u = 0$ on the boundary of D, we take the discrete grid function u^h solving the usual 5-point formula $-(u_N^h + u_S^h + u_E^h + u_W^h - 4u_C^h)/h^2 = f_C$ at the interior grid points and $u^h = 0$ at the grid points on the boundary of D. We denote this equation by $B^h u^h = f^h$. Thus, in this implementation of the algorithm in Equation 2.5, we would approximate the solution r_{n+1} of $Br_{n+1} = R_{n+1}$ by r_{n+1}^h solving $B^h r_{n+1}^h = R_{n+1}^h$. It is clear that, as n increases, u_n^h will converge not to a solution of Equation 3.1 but to a solution to some discrete approximation to Equation 3.1. In fact, the discrete approximation to Equation 3.1 is just the usual 5-point formula derived by variational means. More precisely, since the solution u to Equation 3.1 minimizes $\iint_D \{su_x^2 + tu_y^2 + F(x,y,u)\}dxdy$ where $\frac{\partial F}{\partial u} = f(x,y,u)$, we seek a discrete solution u^h minimizing $Q^h(u^h) \equiv \sum_P s(P)[u_x^h(P)]^2 + \sum_{P'} t(P')[u_y^h(P')]^2 + \sum_{P''} F(P'', u^h(P''))$, where the sums range over those points P at the middle of the horizontal sides of the mesh squares, those points P' at the middle of the vertical sides of the mesh squares, and those points P'' in the interior of D_h, and where u_x^h and u_y^h denote the usual central divided differences at a spacing of $\frac{h}{2}$. This minimization is of course equivalent to solving $J^h(u^h) = 0$, a discrete model of Equation 3.1; if U is any grid function, we can think of $J^h(U) = 0$ as having one equation for each grid point (i,j) in the interior of D_h, namely

$$(3.2)\qquad [s_{i+\frac12,j}(U_{i+1,j} - U_{ij}) - s_{i-\frac12,j}(U_{ij} - U_{i-1,j}) + t_{i,j+\frac12}(U_{i,j+1} - U_{ij})$$
$$- t_{i,j-\frac12}(U_{ij} - U_{ij-2})]/h^2 + f(x_i, y_j, U_{ij}) = 0 \,.$$

Recall that the quadratic form for the discrete Laplacian B^h is the same as Q^h wherein we take $s \equiv t \equiv 1$ and $F \equiv 0$, and note that the quadratic form $(u^h)^T(J_U^{h'})(u^h)$ is just Q^h wherein we take $F(P'', u^h(P'')) = \frac{\partial f}{\partial u}(P'', U(P''))[u^h(P'')]^2$. Since $\frac{\partial f}{\partial u} \geq 0$, it follows easily from the above remarks that the derivatives $J_U^{h'}$ are uniformly B^h-positive definite, independently of h and U. A well-known monotonicity theorem [19, p. 207, for example] tells us that the least eigenvalue of B^h is no less than the least eigenvalue of the discrete Laplacian over a square region enclosing D; we conclude that the smallest eigenvalue of B^h is bounded away from zero as h tends to zero. From this it then follows that the derivatives $J_U^{h'}$ are uniformly B^h-bounded, independently of h and U.

Therefore, the algorithm in Equation 2.5, with J^h and B^h replacing J and U, gives a sequence u_n^h converging at least geometrically to a solution of $J^h(u^h) = 0$, and the geometric convergence bound can be chosen to be, say, λ, independent of the mesh size h. Thus, the number of steps of the algorithm needed to obtain a given accuracy ϵ in solving $J^h(u^h) = 0$ can be bounded by $\frac{\log \epsilon}{\log \lambda}$ independently of h. Of course, the work per step, for example in solving $B^h r^h = R^h$, depends strongly on h. In our numerical examples to come, we choose for D the unit square $(0,1) \times (0,1)$ with the smaller square $\left[\frac{3}{8}, \frac{5}{8}\right] \times \left[\frac{3}{8}, \frac{5}{8}\right]$ deleted from it. As shown by Buzbee et alii [2], the discrete Poisson equations $B^h r^h = R^h$ can be solved very efficiently by direct methods after a certain amount of preprocessing dependent only on the region; in particular, we choose to use the Buneman non-iterative Poisson solver [1] as programmed by J. A. George [9] and kindly provided us by F. W. Dorr.

By using the monotonicity theorem mentioned above, we know that the smallest eigenvalue of B^h is not less than $2\pi^2 + O(h^2)$, the smallest eigenvalue for the 5-point discrete Laplace operator on $(0,1) \times (0,1)$. Thus, on our model region, the convergence rate for the conjugate gradient method is determined by the spectral bounds for $B_h^{-1} J_{u^h}^{h'}$, namely $a^h \geq a$ and $A^h \leq A + \frac{\alpha}{2\pi^2} \approx A + .05\alpha$, where $a \leq s(x,y) \leq A$, $a \leq t(x,y) \leq A$, and $0 \leq \frac{\partial f}{\partial u}(x,y,u) \leq \alpha$. In fact, the actual convergence factor should be no greater than $\dfrac{1 - \sqrt{\frac{a}{A+.05\alpha}}}{1 + \sqrt{\frac{a}{A+.05\alpha}}}$.

4. The Model Problems and Their Solution

As we have indicated above, we shall illustrate our general method by solving the equation

$$(4.1) \qquad (su_x)_x + (tu_y)_y = f(x,y,u)$$

over the domain

$$(4.2) \qquad D = (0,1) \times (0,1) - \left[\frac{3}{8}, \frac{5}{8}\right] \times \left[\frac{3}{8}, \frac{5}{8}\right];$$

although our earlier work assumed for simplicity homogeneous boundary data, we can and do easily consider the more general data

$$(4.3) \qquad u(P) = q(P) \text{ on the boundary of } D.$$

This problem is approximated by the standard 5-point approximation, yielding the system $J^h(u^h) = 0$ as in Equation 3.2, with $u^h(P) = q(P)$ on the boundary of D. The Poisson equation $-\nabla^2 r = R$ with homogeneous Dirichlet data is similarly approximated by the 5-point discrete operator B^h, and equations $B^h r^h = R^h$ are solved by the direct method of Buneman (see [1] and [9]). Ideally we would compute u^h by using the algorithm in Equation 2.5 with J and B replaced by J^h and B^h; to handle the inhomogeneous boundary data of Equation 4.3, we need only choose u_0^h to satisfy that data and then determine r_n and p_n with homogeneous boundary conditions. We now proceed to describe more precisely how this algorithm is coded, in particular how b_n and c_n are determined, how much work is involved in the various stages of the algorithm, *et cetera*.

For each of the test problems and for each value of h the initial guess u_0^h was chosen to be zero inside the region and equal to x^2+y^2 on the boundary.

Our test problems were:

$$(4.4)\quad \begin{cases} [(1+x^2+y^2)u_x]_x + [(1+e^x+e^y)u_y]_y = e^u g(x,y) \\ \text{where} \\ g(x,y) = e^{-(x^2+y^2)}[4 + 6x^2 + 2y^2 + 2e^x + 2(y+1)e^y] \end{cases}$$

and

$$(4.5)\quad \begin{cases} \{[1 + (x-y)^2]u_x\}_x + \{[10 + e^{xy}]u_y\}_y = (1+u)^3 g(x,y) \\ \text{where} \\ g(x,y) = \dfrac{22+6x^2-8xy+2y^2+2e^{xy}(xy+1)}{(1+x^2+y^2)^3} . \end{cases}$$

Both problems have $u(x,y) = x^2 + y^2$ as their unique exact solution.

Initially p_0^h is zero, a scalar γ_{-1} is set to unity, and the starting residual $R_0^h = -J^h(u_0^h)$ is found. Further, a capacitance matrix C^h (see Appendix of [20], and [2]) is constructed and decomposed into Cholesky factors.

On the nth cycle of the conjugate gradient iteration:

(1) A solution r_n^h to the discretized Poisson problem on $(0,1)\times(0,1) - \left(\frac{3}{8},\frac{5}{8}\right)\times\left(\frac{3}{8},\frac{5}{8}\right)$ is found as given in Equations (A.1)-(A.3) of the Appendix of [20];

(2) γ_n is set to the value of $(R_n^h)^T(r_n^h)$ and p_n^h is updated by setting

$$p_n^h = r_n^h + (\gamma_n/\gamma_{n-1})p_{n-1}^h ;$$

(3) A step is taken along p_n^h by finding z_n^h as given in Equation (A.8) of the Appendix of [20] by computing

$$c_n = (p_n^h)^T(R_n^h)/(p_n^h)^T(z_n^h),$$

and by setting

$$u_{n+1}^h = u_n^h + c_n p_n^h .$$

(4) If $\|r_n^h\|$ does not satisfy a threshold test, a new iteration is begun at (1).

The algorithm we use to solve the discrete Poisson equations first performs some preprocessing essentially to eliminate the complications caused by the irregularity of the region over which the differential equation is posed. In the Appendix of [20] we show that the preprocessing costs are proportional to

$$\left(\frac{1}{h}\right)^3 \log_2\left(\frac{1}{h}\right), \tag{4.6}$$

that, once the preprocessing has been done, the Poisson solutions required by step (1) of the conjugate gradient cycle require work proportional to

$$\left(\frac{1}{h}\right)^2 \log_2\left(\frac{1}{h}\right), \tag{4.7}$$

and that the remaining costs per iteration are proportional to

$$\left(\frac{1}{h}\right)^2. \tag{4.8}$$

To see how these costs appear in our examples, refer to Tables 1 and 2. All times listed are in seconds, and computations were carried out on The University of Texas' CDC 6600. The error mentioned in the table is the square root of the sum of the squares of the differences between the true solution of the differential equation and the approximate solution to the difference equation.

TABLE 1

(Problem 4.4)

	Preprocess Time	CG Iteration Time to Reduce Error by Factor h^2	Number of Iterations to Reduce Error by Factor h^2	Per-Iteration Time to Solve Poisson Subproblems	Remaining Per-Iteration Time	Total Solution Time (Sum of Cols. 1 & 2 plus time for initial residual)
$h = \frac{1}{16}$	.394	2.08	6	.141	.206	2.57
$h = \frac{1}{32}$	2.90	9.63	8	.302	.901	12.9
$h = \frac{1}{64}$	23.5	47.2	10	.994	3.72	72.4

TABLE 2

(Problem 4.5)

	Pre-process Time	CG Iteration Time to Reduce Error by Factor h^2	Number of Iterations to Reduce Error by Factor h^2	Per-iteration Time to Solve Poisson Subproblems	Remaining Per-iteration Time	Total Solution Time (Sum of Cols. 1 and 2 Plus Time for Initial Residual)
$h = \frac{1}{16}$	.398	2.58	9	.142	.145	3.05
$h = \frac{1}{32}$	2.82	11.3	12	.305	.638	14.4
$h = \frac{1}{64}$	23.6	54.8	15	.992	2.66	79.5

In the Appendix of [20] the authors consider the conjugate gradient method in brief comparison with an SOR-type process and an ADI-type process, both capable of handling the same sorts of nonlinear elliptic problems. The comparisons are summarized in Table 3.

TABLE 3

	CG Method As Programmed with Direct Poisson Solver	An SOR-type Method	An ADI-type Method
Preprocessing cost is proportional to	$\left(\frac{1}{h}\right)^3 \log_2\left(\frac{1}{h}\right)$	0	0
Number of iterations to reduce error by a factor of h^2 is proportional to	$\log_2\left(\frac{1}{h}\right)$	$\left(\frac{1}{h}\right)\log_2\left(\frac{1}{h}\right)$	$\left[\log_2\left(\frac{1}{h}\right)\right]^2$
Per-iteration cost is proportional to	$\left(\frac{1}{h}\right)^2 \log_2\left(\frac{1}{h}\right)$	$\left(\frac{1}{h}\right)^2$	$\left(\frac{1}{h}\right)^2$
Total cost is proportional to	$\left[\log_2\left(\frac{1}{h}\right)\right]\left[\left(\frac{1}{h}\right)^3 + \left(\frac{1}{h}\right)^2 \log_2\left(\frac{1}{h}\right)\right]$	$\left(\frac{1}{h}\right)^3 \log_2\left(\frac{1}{h}\right)$	$\left(\frac{1}{h}\right)^2\left[\log_2\left(\frac{1}{h}\right)\right]^2$

These estimates were made under the assumption that the SOR-type and ADI-type methods would solve a nonlinear problem over the irregular region we have chosen no more slowly than the usual SOR and ADI methods with optimal parameters would solve the much simpler Poisson problem on the unit square. It is clear that these are terribly optimistic assumptions that could not be expected to hold in general. Therefore the above comparisons of the work needed to solve the problems are very strongly biased in favor of the SOR and ADI methods.

Some further comments are in order regarding the entries in these tables:

1. While the overall cost of the conjugate gradient process is proportional to $\left(\frac{1}{h}\right)^3 \log_2\left(\frac{1}{h}\right)$, it should be observed that the third power is introduced by the preprocessing stage. This is a fixed cost which will be an unbearable proportion of the total solution time only when h is set too small. The preprocessing cost puts a bound on the size of h below which it is uneconomical to use the conjugate gradient method <u>with the Buneman non-iterative Poisson solver</u>. We should remark that this lower bound had not yet been reached for our sample computations. With $h = 1/64$ (approximately 3000 unknowns) the preprocessing time was only about 30% of the total time (about 47% of the time required for the conjugate gradient iterations).

2. The conjugate gradient method could be programmed with any Poisson solver for the chosen region. In particular an ADI or Finite Element Poisson solver for the irregular region, if "well-tuned," would convert the method we propose into one whose cost would be proportional only to $\left(\frac{1}{h}\right)^2$ times possibly a power of $\log_2\left(\frac{1}{h}\right)$.

3. If the Buneman method is chosen to solve the Poisson subproblems, it should be noted that the preprocessing time depends only upon the geometry of the region on which the nonlinear problem is given and the value of h chosen for the discretization. If many problems are to be solved for the same geometry, the preprocessing need not be carried out more than once. Each problem will require only an expenditure proportional to $\left(\frac{1}{h}\right)^2 \log_2\left(\frac{1}{h}\right)$ for its solution even with the Buneman Poisson solver.

4. The conjugate gradient method is parameter-free. If a good Poisson solver is available, then the conjugate gradient method, unlike SOR or ADI, gives its optimal performance without fine-tuning or elaborate parameter estimation.

5. As shown in the Appendix of [20], each step of the conjugate gradient method requires only a small percentage of work beyond that required by SOR or ADI methods; thus the constants of proportionality appearing in our estimates of the work per iteration will be essentially the same for the various methods.

Acknowledgements. The authors wish to thank F. W. Dorr, G. H. Golub, and D. M. Young for their helpful discussions related to this problem.

References

1. Buneman, O. "A compact non-iterative Poisson-solver," Rep. SU-IPR-294, Inst. Plasma Res., Stanford Univ., Stanford, Calif. (1969).

2. Buzbee, B. L., Dorr, F. W., George, J. A., and Golub, G. H. "The direct solution of the discrete Poisson equation on irregular regions," SIAM J. Num. Anal. 8, 722-736 (1971).

3. Buzbee, B. L., Golub, G. H., and Nielson, C. W. "On direct methods for solving Poisson's equations," SIAM J. Num. Anal. 7, 627-656 (1970).

4. Daniel, J. W. "The conjugate gradient method for linear and nonlinear operator equations," Ph.D. thesis, Mathematics, Stanford University (1965).

5. Daniel, J. W. "The conjugate gradient method for linear and nonlinear operator equations," SIAM J. Num. Anal. 4, 10-26 (1967).

6. Dorr, F. W. "The direct solution of the discrete Poisson equation on a rectangle," SIAM Rev. 12, 248-263 (1970).

7. Engeli, M., Ginsburg, Th., Rutishauser, H., and Stiefel, E. "Refined iterative methods for the computation of the solution and the eigenvalues of self-adjoint boundary value problems," Mitt. Inst. Agnew. Math. Zurich, No. 8 (1959).

8. Fletcher, R., and Reeves, C. M. "Function minimization by conjugate gradients," Comput. J. 7, 149-154 (1964).

9. George, J. A. "The use of direct methods for the solution of the discrete Poisson equation on non-rectangular regions," Rep. STAN-CS-70-159, Computer Sci. Dept., Stanford Univ., Stanford, Calif. (1970).

10. Hayes, R. M. "Iterative methods of solving linear problems in Hilbert space," in Contributions to the Solution of Systems of Linear Equations and the Determination of Eigenvalues, O. Taussky (ed.), Nat. Bur. Standards Appl. Math. Ser. 39, 71-104 (1954).

11. Hestenes, M. R. "The conjugate gradient method for solving linear systems," Proc. Symp. Appl. Math. VI, Num. Anal., 83-102 (1956).

12. Hestenes, M. R., and Stiefel, E. "Method of conjugate gradients for solving linear systems," J. Res. Nat. Bur. Standards 49, 409-436 (1952).

13. Lanczos, C. "An iteration method for the solution of the eigenvalue problem of linear differential and integral operators," J. Res. Nat. Bur. Standards 45, 255-283 (1950).

14. Reid, J. K. "On the method of conjugate gradients for the solution of large sparse systems of linear equations," in Proc. Conf. on Large Sparse Sets of Linear Equations, Academic Press, New York (1971).

15. Reid, J. K. "The use of conjugate gradients for systems of linear equations possessing 'Property A,'" SIAM J. Num. Anal. 9, 325-332 (1972).

16. Stiefel, E. "Uber einige methoden der Relaxationsrechnung," ZAMP 3, 1-33 (1952).

17. Stiefel, E. "Recent developments in relaxation techniques," Proc. Inter. Cong. Appl. Math. I, 384-391 (1954).

18. Stiefel, E. "Relaxationsmethoden bester Strategie yur Losung linearer Gleichungs-systeme," Comment. Math. Helv. 29, 157-179 (1955).

19. Young, David M. Iterative Solution of Large Linear Systems, Academic Press, (1971).

20. Bartels, Richard, and Daniel, James W. "A conjugate gradient approach to non-linear elliptic boundary value problems in irregular regions," CNA-63, Center for Numerical Analysis, Univ. of Texas, Austin, Texas (1973).

GOOD APPROXIMATION BY SPLINES WITH VARIABLE KNOTS. II

Carl de Boor

This note is concerned with a rather simple algorithm for the adaptive placement of break points when approximating a function g, given *implicitly* by some ordinary differential equation, by piecewise polynomial (pp) functions. The algorithm is suggested by the methods and results of an analysis of the distance of a given function from pp functions with a fixed number of pieces of fixed order as carried out by Rice [7], Phillips [6], McClure [5], Burchard [3], de Boor [1] and Dodson [4], and is a variant of the algorithm used by Dodson. An extension to the numerical solution of certain partial differential equations has been made by Sewell [9].

1. COLLOCATION

Specifically, assume that g is known to satisfy

$$(D^m g)(t) = F(t; g(t), \ldots, (D^{m-1}g)(t)) \quad \text{on} \quad [a,b], \tag{1}$$
$$\beta_i g = c_i, \qquad i = 1, \ldots, m$$

for some known real valued function $F = F(t; z_o, \ldots, z_{m-1})$ on $\mathbb{R}^{m+1}$, some known continuous linear functionals $\beta_1, \ldots, \beta_m$ on $C^{(m-1)}[a,b]$, and some known real numbers $c_1, \ldots, c_m$.

The approximation is to be chosen from the class of functions

$$S_\Delta := \mathbb{P}_{k,\Delta} \cap C^{(m-1)}[a,b]$$

with $\Delta := (t_i)_1^{N+1}$ a (strict) partition for $[a,b]$, i.e.,

$$a = t_1 < t_2 < \cdots < t_{N+1} = b$$

and $\mathbb{P}_{k,\Delta}$ the class of all piecewise polynomial functions of order k with break point sequence Δ. More explicitly, $\mathbb{P}_{k,\Delta}$ consists of those functions on

Sponsored by the United States Army under Contract No. DA-31-124-ARO-D-462.

$[a,b]$ which, on each interval (t_i, t_{i+1}), coincide with some polynomial of degree $< k$. (We choose not to distinguish between two such pp functions if they differ only on the points of Δ.)

The particular approximation for g is to be chosen from S_Δ by collocation. With

$$-1 \le \rho_1 < \rho_2 < \cdots < \rho_{k-m} \le 1$$

given, set

$$\tau_{(i-1)(k-m)+r} := [t_{i+1} + t_i + \rho_r(t_{i+1} - t_i)]/2, \quad r = 1,\ldots,k-m, \quad i = 1,\ldots,N.$$

Then we look for $f \in S_\Delta$ such that

$$(1') \quad \begin{aligned} &(D^m f)(\tau_i) = F(\tau_i; f(\tau_i),\ldots,(D^{m-1}f)(\tau_i)), \quad i = 1,\ldots,N(k-m), \\ &\beta_i f = c_i, \quad i = 1,\ldots,m. \end{aligned}$$

The following facts (and others) about this approximation process are described more explicitly (and proved) in [2]. The first two points are already in Russell and Shampine's paper [8] provided $-1 = \rho_1$ and $\rho_{k-m} = 1$. The third point is a consequence of Lemma 3.1 of [2], esp. (3.9). The last point can be established by an extension of the argument for Theorem 4.1 in [2].

Assume that g is an isolated solution of (1), F is sufficiently smooth, and (β_i) satisfies a certain condition. Then

(i) for every Δ with

$$|\Delta| := \max_i \Delta t_i$$

sufficiently small, (1') has exactly one solution "near" g.

Call this solution f_Δ.

(ii) The error estimate

$$\|D^j(g - f_\Delta)\|_\infty \le \text{const}_k |\Delta|^{k-m} \|D^k g\|_\infty$$

holds for $j = 0,\ldots,m$, for some constant depending on F, (β_i), (c_i) and (ρ_i).

(iii) With the abbreviation

$$\|h\|_{(i)} := \sup_{t_i \le t \le t_{i+1}} |h(t)| ,$$

we have, more explicitly,

(2) $$\|g - f_\Delta\|_\infty \le \text{const} \max_i |\Delta t_i|^{k-m} \|D^k g\|_{(i)} + o(|\Delta|^{k-m}) .$$

Assume, in addition, that $\rho_1, \ldots, \rho_{k-m}$ *are the zeros of the* $(k-m)$*th Legendre polynomial. Then*

(iv) $$\|D^j(g - f_\Delta)\|_\infty = 0(|\Delta|^{k-j}) , \quad j = 0, \ldots, m ,$$

and, for $i = 1, \ldots, N+1$, *even*

$$|D^j(g - f_\Delta)(t_i)| = 0(|\Delta|^{2(k-m)}) , \quad j = 0, \ldots, m-1 .$$

(v) *Also, if* $2m < k$, *then for some constant and for* $i = 1, \ldots, N$,

(3) $$\|g - f_\Delta\|_{(i)} \le \text{const} |\Delta t_i|^k \|D^k g\|_{(i)} + o(|\Delta|^k) .$$

2. THE KNOT PLACEMENT ALGORITHM

The bound (3) above implies that, for sufficiently small $|\Delta|$,

$$\|g - f_\Delta\| \le \text{const} \max_i |\Delta t_i|^k \|D^k g\|_{(i)}$$

and therefore suggests that the break points $t_2, \ldots, t_N$ be placed so as to minimize

(4) $$\max_i |\Delta t_i|^k \|D^k g\|_{(i)} .$$

Since

$$s(\alpha, \beta) := |\beta - \alpha|^k \sup_{\alpha < t < \beta} |(D^k g)(t)|$$

is a continuous function of α and β (if $D^k g$ is continuous) and monotone, increasing in β and decreasing in α , (4) is minimized when $t_2, \ldots, t_N$ are so chosen that

$$|\Delta t_i|^k \| D^k g \|_{(i)} = \text{constant for } i = 1, \ldots, N .$$

The exact determination of such $t_2, \ldots, t_N$ is a somewhat expensive task, and is impossible anyway if, as in our case, $D^k g$ is not known. But, the task is obviously equivalent to determining $t_2, \ldots, t_N$ so that

$$\Delta t_i \| |D^k g|^{1/k} \|_{(i)} = \text{constant for } i = 1, \ldots, N ,$$

and produces therefore asymptotically the same distribution of t_i's as the problem of determining $t_2, \ldots, t_N$ so that

$$\int_{t_i}^{t_{i+1}} |(D^k g)(t)|^{1/k} dt = \frac{1}{N} \int_a^b |(D^k g)(t)|^{1/k} dt , \quad i = 1, \ldots, N .$$

This latter problem is very easy to solve if we replace $|D^k g|$ by a piecewise constant approximation

$$h \approx |D^k g| .$$

For then,

$$I(t) := \int_a^t (h(s))^{1/k} ds$$

is an easily computable continuous and monotone increasing piecewise linear function, hence so is I^{-1} (ignoring the possibility of multivaluedness at some break points), and the stated task amounts to nothing more than evaluating the known piecewise linear function I^{-1} at the $N-1$ points

$$i\, I(b)/N , \quad i = 1, \ldots, N-1 .$$

It remains to discuss how one might obtain a piecewise constant approximation h to $|D^k g|$. With Δ the current break point sequence, and f_Δ the resulting collocation approximation to g, one procedure consists in determining

$$H(t) := \int_a^t h(s) ds \;\varepsilon\; \mathbb{P}_{2,\Delta} \cap C$$

so as to approximate the increasing piecewise constant function

$$\mathrm{Var}_{[a,t]} D^{k-1} f_\Delta \approx \mathrm{Var}_{[a,t]} D^{k-1} g = \int_a^t |(D^k g)(s)|\, ds \ .$$

Specifically, choose $h \in \mathbb{P}_{1,\Delta}$ so that

$$\text{(5a)} \qquad h(t) = \begin{cases} 2\,\dfrac{|\Delta f_{3/2}|}{t_3 - t_1} & , \text{ on } (t_1, t_2) \\[2ex] \dfrac{|\Delta f_{i-1/2}|}{t_{i+1} - t_{i-1}} + \dfrac{|\Delta f_{i+1/2}|}{t_{i+2} - t_i} & , \text{ on } (t_i, t_{i+1}) \ , \quad i = 2, \ldots, N-1 \ , \\[2ex] 2\,\dfrac{|\Delta f_{N-1/2}|}{t_{N+1} - t_{N-1}} & , \text{ on } (t_N, t_{N+1}) \end{cases}$$

with

$$\text{(5b)} \qquad f_{i+1/2} := D^{k-1} f_\Delta \ \text{ on } (t_i, t_{i+1}) \ , \quad \text{all } i \ .$$

This amounts to taking for h on (t_i, t_{i+1}) the slope at $t_{i+1/2}$ of the parabola which interpolates $\mathrm{Var}_{[a,t]} D^{k-1} f_\Delta$ at $t_{i-1/2}$, $t_{i+1/2}$, and $t_{i+3/2}$, with $t_{j+1/2} := (t_j + t_{j+1})/2$. The particular choices in the first and the last interval made here were suggested to me by Robert E. Lynch.

I have used this algorithm in an iterative procedure as follows:

Given an initial break point sequence Δ and an initial guess $\hat{f}$ for g.

(i) Starting with the current guess $\hat{f}$, use Newton's method or some other iterative scheme to obtain the solution f_Δ of (1') "near" g.

(ii) If a better approximation to g is desired, then, from f_Δ, obtain $h \in \mathbb{P}_{1,\Delta}$ as in (5a-b), increment N, and determine

$$t_{i+1} = I^{-1}(iI(b)/N) \ , \qquad i = 1, \ldots, N-1$$

with $I(t) = \int_a^t (h(s))^{1/k} ds$. Then set $\hat{f} := f_\Delta$ and go back to (i).

3. COMMENTS

There does not exist at present any compelling reason for preferring the above knot placement algorithm to any other. The algorithm takes few operations and has given very satisfactory results in many instances. The algorithm has failed badly in situations when the o-term in (3) could not be ignored. Typically, the algorithm failed to put any break points into a relatively large subinterval on which g (and therefore the computing f_Δ) was essentially constant compared to its behaviour on the rest of $[a,b]$. The next computed f_Δ then turned out to be a much poorer approximation because now the larger $|\Delta|$ caused the global term $o(|\Delta|^k)$ in (3) to overshadow the local term $\max_i |\Delta t_i|^k \|D^k g\|_{(i)}$ for whose minimization the new Δ was presumably selected correctly. This difficulty can be alleviated merely by modifying the algorithm so as to prevent a drastic increase in $|\Delta|$ as N increases.

My repeated attempts to analyze rigorously the effectiveness of this and other knot placement algorithms have so far come to grief because of the following facts:

(i) No matter how cleverly one approximates g by elements of

$$\mathbb{P}_{k,N} := \text{pp functions of order } k \text{ on } [a,b] \text{ consisting of } N \text{ pieces,}$$

the order of convergence will never be better than N^{-k} (aside from trivial exceptions). Specifically [6],

$$\lim_{N\to\infty} N^k \operatorname{dist}_\infty(g, \mathbb{P}_{k,N}) = 2^{1-2k}/k!\ \|D^k g\|_{1/k} .$$

(ii) No matter how poorly one selects the function r on $[a,b]$ with

$$\inf_{a<t<b} r(t) > 0 ,$$

if $\Delta = (t_i)_1^{N+1}$ is so chosen that

$$\int_{t_i}^{t_{i+1}} r(s)ds = \text{constant} , \qquad i = 1,\ldots,N ,$$

then the order of convergence of the collocation approximation is still N^{-k}, i.e.,

$$\text{(6)} \qquad \|g - f_\Delta\|_\infty \leq \text{const } N^{-k}$$

with the constant depending on $\inf r(t)$ and the collocation process.

It follows that, in the typical (asymptotic) error estimates of the form (6), any two knot placement algorithms will be distinguished only by the value of the constant, and estimating these constants is very difficult.

4. OTHER KNOT PLACEMENT ALGORITHMS

To the extent that the numerical solution of an ordinary differential equation by finite differences can be interpreted as approximation by pp functions, with the mesh points the break points, the adaptive placement of break points is a very old idea indeed; it is practised routinely in the numerical solution of initial value problems.

In the numerical solution of boundary value problems by finite difference techniques, H. B. Keller has used for some time quite successfully the scheme of placing the mesh points so as to equalize the variation over subintervals, i.e., so that

$$\int_{t_i}^{t_{i+1}} |(Dg)(s)|\,ds = \text{constant for all } i ,$$

replacing, I imagine, Dg by some piecewise constant approximation to it obtained from the current approximation to g. This corresponds to minimizing

$$\max_i \Delta t_i \, \|Dg\|_{(i)}$$

and is therefore based on the assumption that the error in the approximation is essentially proportional to $\text{dist}_\infty(g, \mathbb{P}_{1,\Delta})$. But if the error in the approximation behaves like

$$\max_i (\Delta t_i)^2 \|D^4 g\|_{(i)} ,$$

as seems to be the case for the standard finite difference approximation to a second order two point boundary value problem, then one would expect to obtain a better distribution from choosing Δ so that

$$\int_{t}^{t_{i+1}} |D^4 g|^{1/2} = \text{constant for all } i .$$

In a similar vein, if the collocation points $\rho_1, \ldots, \rho_{k-m}$ are chosen uniformly spaced, then the error estimate (3) fails to hold. Instead, (2) turns out to describe the error behaviour quite accurately, so that now Δ should be chosen so as to minimize

$$\max_i |\Delta t_i|^{k-m} \| D^k g \|_{(i)} .$$

In conclusion, break point or mesh point placement schemes based on the (guessed) local behaviour of the unknown function g seem to be justified only if (i) the error of the approximation depends on the local behaviour of g, and if (ii) the placement scheme is based on exactly how the approximation error depends on the local behaviour of g. This statement is, of course, not invalidated by the fact that any given scheme performs quite satisfactorily in many instances where such justification cannot be given.

REFERENCES

1. C. de Boor, Good approximation by splines with variable knots, in "Spline functions and approximation theory", A. Meir and A. Sharma ed., ISNM Vol. 21, Birkhäuser Verlag, Basel (1973), 57-72.
2. C. de Boor and B. Swartz, Collocation at Gaussian points, SIAM J. Numer. Anal. 10 (1973).
3. H. Burchard, Splines (with optimal joints) are better, to appear in J. Applicable Math. 1.
4. D. S. Dodson, Optimal order approximation by polynomial spline functions, Ph.D. Thesis, Purdue Univ., Lafayette, Ind., Aug. 1972.
5. D. E. McClure, Feature selection for the analysis of line patterns, Ph.D. Thesis, Brown Univ., Providence, R. I., 1970.
6. G. M. Phillips, Error estimates for best polynomial approximation, in "Approximation Theory", A. Talbot ed., Academic Press, London (1970), 1-6.
7. J. R. Rice, On the degree of convergence of nonlinear spline approximation, in "Approximations with special emphasis on spline functions", I. J. Schoenberg ed., Academic Press, New York (1969), 349-365.
8. R. D. Russell and L. F. Shampine, A collocation method for boundary value problems, Numer. Math. 19 (1972), 1-28.

9. E. G. Sewell, Automatic generation of triangulations for piecewise polynomial approximation, Ph. D. Thesis, Purdue Univ., Lafayette, Ind., Dec. 1972.

CONFORMING AND NONCONFORMING
FINITE ELEMENT METHODS FOR SOLVING THE PLATE PROBLEM

P.G. Ciarlet

Abstract. For the clamped plate problem ($\Omega \subset R^2$):

$$\Delta^2 u = f \text{ in } \Omega,\ u = \frac{\partial u}{\partial n} = 0 \text{ on } \partial\Omega,$$

various finite element methods are described and compared as regards their asymptotic order of convergence. A particular emphasis is put upon the connections between the patch test and convergence for nonconforming methods.

1. THE PLATE PROBLEM

In what follows, Ω is a bounded open subset of the plane, with boundary $\partial\Omega$. For a given integer $m \geq 0$, we let

$$|v|_{m,\Omega} = \Big(\sum_{|\alpha|=m} \int_\Omega |\partial^\alpha v|^2 dxdy\Big)^{1/2},\ \|v\|_{m,\Omega} = \Big(\sum_{\ell=0}^{m} |v|^2_{\ell,\Omega}\Big)^{1/2}.$$

Over the Sobolev space $V = H^2_0(\Omega)$, the semi-norm $|\cdot|_{2,\Omega}$ is a norm, which is equivalent to the norm $\|\cdot\| = \|\cdot\|_{2,\Omega}$.

Given a function $f \in L^2(\Omega)$, consider the problem : Find a function $u \in V$ which satisfies

$$a(u,v) = (f,v) \text{ for all } v \in V, \tag{1.1}$$

where $(\cdot,\cdot)$ denotes the scalar product in the space $L^2(\Omega)$ and the bilinear form $a(\cdot,\cdot)$ is defined by

(1.2) $$a(u,v) = \int_\Omega \{\Delta u\, \Delta v + (1-\sigma)\left(2\frac{\partial^2 u}{\partial x \partial y}\frac{\partial^2 v}{\partial x \partial y} - \frac{\partial^2 u}{\partial x^2}\frac{\partial^2 v}{\partial y^2} - \frac{\partial^2 u}{\partial y^2}\frac{\partial^2 v}{\partial x^2}\right)\}dxdy.$$

This is the variational formulation of the plate problem, as may be found in Landau & Lifchitz [16, Chapter 2]. The constant σ is the Poisson's coefficient of the plate and it satisfies the inequalities $0 < \sigma < \frac{1}{2}$. Therefore, the bilinear form $a(\cdot,\cdot)$ is V-elliptic, since

$$a(v,v) = \sigma|\Delta v|^2_{0,\Omega} + (1-\sigma)|v|^2_{2,\Omega} \text{ for all } v \in V.$$

Since it is also continuous over the space V, the variational problem (1.1) has a unique solution $u \in V$.

Using the Green's formulas (with standard notations) :

(1.3) $$\int_\Omega \Delta u\, \Delta v\, dxdy = \int_\Omega \Delta^2 u\, v\, dxdy - \oint_{\partial\Omega} \frac{\partial \Delta u}{\partial n} v\, d\gamma + \oint_{\partial\Omega} \Delta u \frac{\partial v}{\partial n} d\gamma,$$

(1.4) $$\int_\Omega \{2\frac{\partial^2 u}{\partial x \partial y}\frac{\partial^2 v}{\partial x \partial y} - \frac{\partial^2 u}{\partial x^2}\frac{\partial^2 v}{\partial y^2} - \frac{\partial^2 u}{\partial y^2}\frac{\partial^2 v}{\partial x^2}\}dxdy = \oint_{\partial\Omega} \{-\frac{\partial^2 u}{\partial t^2}\frac{\partial v}{\partial n} + \frac{\partial^2 u}{\partial n \partial t}\frac{\partial v}{\partial t}\}d\gamma,$$

it is seen that if the solution u of (1.1) is smooth enough, then it is also the solution of the problem

(1.5) $$\Delta^2 u = f \text{ in } \Omega,$$

(1.6) $$u = \frac{\partial u}{\partial n} = 0 \text{ on } \partial\Omega,$$

which is the simplest mathematical model for a clamped plate.

Finally, it is known that the function u is in the space $H^3(\Omega) \cap \overset{\circ}{H}{}^2(\Omega)$ if $\bar{\Omega}$ is a convex polygon, an assumption often satisfied by plates.

2. CONFORMING METHODS

The bilinear form $a(\cdot,\cdot)$ being continuous and V-elliptic, there exist two constants M and $\alpha > 0$ such that

(2.1) $$|a(u,v)| \leq M \|u\| \|v\| \text{ for all } u,v \in V,$$

(2.2) $$\alpha \| u \|^2 \leq a(u,u) \text{ for all } u \in V.$$

Given a finite-dimensional subspace V_h of the space V, the discrete analogue of problem (1.1) consists in finding a function $u_h \in V_h$ such that

(2.3) $$a(u_h,v_h) = (f,v_h) \text{ for all } v_h \in V_h.$$

This problem has a unique solution u_h in view of inequality (2.2). Using the inequalities

$$\alpha \| u-u_h \|^2 \leq a(u-u_h,u-u_h) = a(u-u_h,u-v_h) \leq M \| u-u_h \| \, \| u-v_h \|$$

valid for any function $v_h \in V_h$, we deduce the inequality

(2.4) $$\| u-u_h \| \leq C \inf_{v_h \in V_h} \| u-v_h \|,$$

where $C = \frac{M}{\alpha}$ is a constant independent of the subspace V_h. Therefore the problem of estimating the error $\| u-u_h \|$ is reduced to a problem in approximation theory : evaluate $\inf_{v_h \in V_h} \| u-v_h \|$.

To obtain inequality (2.4), an essential use has been made of the inclusion $V_h \subset V$: in this case we shall say that the associated discrete method, and the finite elements themselves, are conforming.

Assuming the set $\bar{\Omega}$ is a polygon, we may establish a triangulation $\mathcal{C}_h$ over $\bar{\Omega}$, i.e., we write $\bar{\Omega} = \bigcup_{K \in \mathcal{C}_h} K$, where the finite elements K are triangles (or rectangles; cf. Table 2) with the usual assumptions that their interiors are pairwise disjoint, and that any side of a triangle is either a side of another triangle or a subset of the boundary $\partial\Omega$. We let

(2.5) $$h_K = \text{diam}(K) \text{ for all } K \in \mathcal{C}_h, \text{ and } h = \max_{K \in \mathcal{C}_h} h_K.$$

Also, over each triangle K, we are given a finite-dimensional space P_K of functions. Then with the triangulation $\mathcal{C}_h$ and the spaces P_K is associated a space V_h of functions $v_h : \bar{\Omega} \to R$ in such a way that

$$v_h|_K \in P_K \text{ for all } K \in \mathcal{C}_h.$$

Since the functions of the spaces P_K are smooth in general (cf. the examples), the inclusion $V_h \subset H^2(\Omega)$ will be in turn a consequence of the simpler inclusion $V_h \subset \mathcal{C}^1(\bar{\Omega})$, which is itself a consequence of an appropriate choice of the degrees of freedom in each triangle. As a consequence of this inclusion, the boundary conditions (1.6) can be exactly satisfied (cf. the examples).

Let P_k denotes the space of all polynomials of degree $\leq k$ in the variables x and y. If the fundamental inclusion

(2.6) $$P_k \subset P_K$$

holds-together with additional, but somehow less important, conditions such as that the elements should not become "flat" in the limit-, then the local error bounds

(2.7) $$|u-\Pi_h u|_K|_{m,K} \leq C \, |u|_{k+1,K} \, h_K^{k+1-m}, \quad 0 \leq m \leq k,$$

hold, where $\Pi_h u$ is the V_h-interpolate of the function u. It follows from inequalities

(2.7) that

$$\inf_{v_h \in V_h} |u-v_h|_{2,\Omega} \leq C\ |u|_{k+1,\Omega} h^{k-1}, \tag{2.8}$$

and therefore, we obtain an $O(h)$ convergence for conforming methods if the solution u lies in the space $H^3(\Omega)$ and if the "minimal" inclusion

$$P_2 \subset P_K \tag{2.9}$$

holds. For examples of subspaces satisfying inequalities (2.7), see for instance Babuška & Aziz [2], Bramble & Hilbert [6], Ciarlet [8], Ciarlet and Raviart [9] Raviart [20], Strang [22], Strang & Fix [24].

Some examples of conforming finite elements are given in Table 1, for each of which we give the definition and dimension of the space P_K together with the "best" inclusion of the form (2.6), the order of convergence in the norm $\|\cdot\| = \|\cdot\|_{2,\Omega}$, references regarding the origin of the element and references where proofs of local error bounds such as those of (2.7) may be found. We are using the following conventions for the degrees of freedom :

- $\bullet$ u.
- $\odot$ $\partial^\alpha u$, $|\alpha| = 0,1$.
- ◎ $\partial^\alpha u$, $|\alpha| = 0,1,2$.
- —┼— normal derivative at mid-point of side.

We could have also included the rectangular element of Bogner, Fox & Schmit [5], the composite quadrilateral element of Sander [21] and Fraeijs de Veubeke [12] for which Ciavaldini and Nédélec (to appear) have recently given error bounds, etc.

3. NONCONFORMING METHODS

For various reasons, it is desirable to consider spaces V_h similar to those of Section 2, but which now violate the inclusion $V_h \subset V$. This situation corresponds to what we shall call nonconforming methods (or finite elements). For such methods, the inclusion $P_2 \subset P_K$ will now be only one of the conditions for obtaining convergence, the other one being the famous patch test (as we shall see later on in this Section), empirically devised by B.M. Irons (see for instance Irons & Razzaque [14] for a pertinent discussion). The patch test has first been studied from a mathematical standpoint by G. Strang [23,24], and then by Crouzeix & Raviart [11], and their ideas are also used in a forthcoming paper of P. Lascaux and P. Lesaint, where complete proofs of convergence are given for the three examples which we consider in the sequel.

What might be new here is an attempt to give an a priori justification for the patch test, the essential tool being a lemma about bilinear forms vanishing over polynomial subspaces, which is similar to the Bramble-Hilbert lemma for linear forms (cf. [6]). These ideas will be more fully developed in a forthcoming paper by P.-A. Raviart and the author.

Since the functions of the space V_h are smooth over each finite element K, the expression

TABLE 1

Some conforming finite elements for the plate problem.

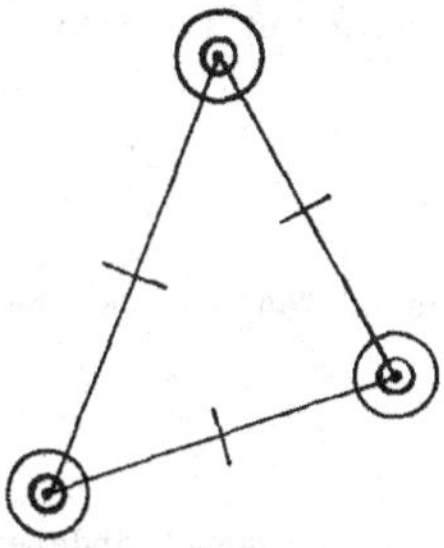	$P_K = P_5$. $\dim P_K = 21$. $\Vert u-u_h \Vert = O(h^4)$ if $u \in H^6(\Omega)$. Origin : Zlámal [26] ; see page 209 of Zienkiewicz [25]. Local error bounds : Zlámal [26].
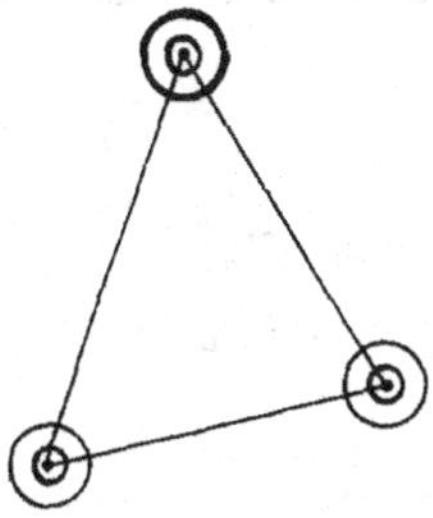	$P_K = \{p \in P_5;\ \frac{\partial p}{\partial n}$ is a pol. of degree ≤ 3 along each side$\}$. $\dim P_K = 18$; $P_4 \subset P_K$. $\Vert u-u_h \Vert = O(h^3)$ if $u \in H^5(\Omega)$. Origin : see page 209 of Zienkiewicz [25]. Local error bounds : Bramble & Zlámal [7].
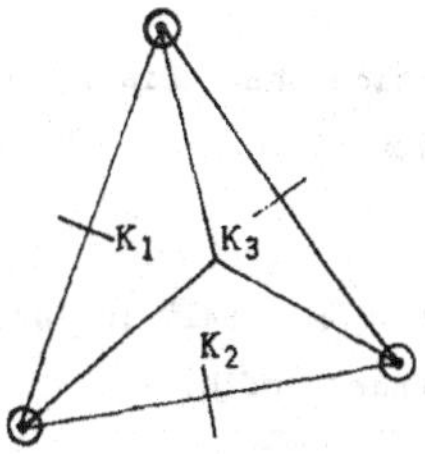 $K = \bigcup_{i=1}^{3} K_i$	$P_K = \{p \in \mathcal{C}^1(K),\ p\vert_{K_i} \in P_3,\ 1 \leq i \leq 3\}$. $\dim P_K = 12$; $P_3 \subset P_K$. $\Vert u-u_h \Vert = O(h^2)$ if $u \in H^4(\Omega)$. Origin : Clough & Tocher [10]. Local error bounds : Ciarlet (to appear).

$$(3.1) \qquad a_h(u_h,v_h) = \sum_{K\in\mathcal{E}_h} \int_K \{\ldots\}dxdy$$

makes sense for all functions $u_h, v_h \in V_h$, the integrand being the same as in (1.2). Thus the natural way of defining the discrete problem is to find $u_h \in V_h$ such that

$$(3.2) \qquad a_h(u_h,v_h) = (f,v_h) \text{ for all } v_h \in V_h,$$

assuming that the inclusion $V_h \subset L^2(\Omega)$ holds. In addition, it turns out (but this fact needs a proof) that the expression

$$(3.3) \qquad \|v_h\|_h = \Big(\sum_{K\in\mathcal{E}_h} |v_h|^2_{2,K}\Big)^{1/2}$$

is a norm over the spaces V_h associated with the three examples of Table 2, so that the bilinear form $a_h(\cdot,\cdot)$ is <u>uniformly</u> V_h-elliptic since

$$(3.4) \qquad \alpha\|u_h\|^2_h \leq a_h(u_h,u_h) \text{ for all } u_h \in V_h,$$

with $\alpha = 1-\sigma$. Also, it is easily established that there exists a constant M independent of h such that

$$(3.5) \qquad |a_h(u_h,v_h)| \leq M\|u_h\|_h\|v_h\|_h \quad \text{for all } u_h,v_h \in V_h.$$

Given an arbitrary element $v_h = u_h - w_h$ in the space V_h, we may write, using (1.1), (3.2), (3.4) and (3.5) :

$$\begin{aligned}\alpha\|u_h-v_h\|^2_h &\leq a_h(u_h-v_h,u_h-v_h)\\ &= a_h(u-v_h,u_h-v_h) + (f,u_h-v_h) - a_h(u,u_h-v_h)\\ &\leq M\|u-v_h\|_h\|u_h-v_h\|_h + |(f,w_h) - a_h(u,w_h)|,\end{aligned}$$

and thus,

$$\|u_h-v_h\|_h \leq \frac{M}{\alpha}\|u-v_h\|_h + \frac{|(f,w_h)-a_h(u,w_h)|}{\|w_h\|_h}$$

from which we deduce the inequality

$$(3.6) \qquad \|u-u_h\|_h \leq C\Big(\inf_{v_h\in V_h}\|u-v_h\|_h + \sup_{w_h\in V_h}\frac{|(f,w_h)-a_h(u,w_h)|}{\|w_h\|_h}\Big),$$

where the constant $C = \max\{1+\frac{M}{\alpha}, \frac{1}{\alpha}\}$ is independent of h. Notice that this inequality generalizes the inequality of (2.4), since the expression

$$(3.7) \qquad E_h(u,w_h) = (f,w_h)-a_h(u,w_h)$$

would be zero in the case of a conforming method. Assuming the "minimal" inclusion $P_2 \subset P_K$, we see that (C denotes as usual any constant independent of h)

$$(3.8) \qquad \inf_{v_h\in V_h}\|u-v_h\|_h \leq C\,|u|_{3,\Omega}h$$

and therefore, in view of inequality (3.6), we need to prove an inequality such as

$$(3.9) \qquad |E_h(u,w_h)| \leq C\,h|u|_{3,\Omega}\|w_h\|_h \text{ for all } u \in H^3(\Omega),\ w_h \in V_h.$$

It follows from Green's formulas (1.3) and (1.4) that

(3.10) $$E_h(u,w_h) = E_h^0(u,w_h) + E_h^1(u,w_h),$$

with

(3.11) $$E_h^0(u,w_h) = \sum_{K\in\mathscr{E}_h} \oint_{\partial K} \{\frac{\partial\Delta u}{\partial n} w_h - (1-\sigma)\frac{\partial^2 u}{\partial n\partial t}\frac{\partial w_h}{\partial t}\} d\gamma,$$

(3.12) $$E_h^1(u,w_h) = \sum_{K\in\mathscr{E}_h} \oint_{\partial K} \left(-\Delta u + (1-\sigma)\frac{\partial^2 u}{\partial t^2}\right) \frac{\partial w_h}{\partial n}\, d\gamma.$$

Rather than giving a general discussion, we shall henceforth concentrate on one specific example : the Adini's rectangle (cf. Table 2), which may be used if the set $\bar\Omega$ is a rectangle. Although some simplifications will indeed result from this choice, all the basic ideas are present.

First, we notice that the term $E_h^0(u,w_h)$ is identically zero : this is a consequence of the inclusion $V_h \subset \mathscr{C}^0(\bar\Omega)$ and of the boundary condition $v_h = 0$ on $\partial\Omega$ which may be exactly satisfied by the functions v_h of V_h.

If we denote by n_x and n_y the components of the outward normal vector along the boundary of one Adini's rectangle, we may write $E_h^1(u,w_h) = E_h^{1,x}(u,w_h) + E_h^{1,y}(u,w_h)$ with

(3.13) $$E_h^{1,x}(u,w_h) = \sum_{K\in\mathscr{E}_h} \oint_{\partial K} \left(-\Delta u + (1-\sigma)\frac{\partial^2 u}{\partial t^2}\right)\frac{\partial w_h}{\partial x} n_x\, d\gamma$$

and with a similar expression for $E_h^{1,y}$. Taking into account the second boundary condition $\frac{\partial w_h}{\partial n} = 0$ verified at all the nodes along $\partial\Omega$, we obtain

(3.14) $$E_h^{1,x}(u,w_h) = \sum_{K\in\mathscr{E}_h} E_{h,K}^{1,x}\left(u,\frac{\partial w_h}{\partial x}\right)$$

with

(3.15) $$E_{h,K}^{1,x}\left(u,\frac{\partial w_h}{\partial x}\right) = \oint_{\partial K} \left(-\Delta u + (1-\sigma)\frac{\partial^2 u}{\partial t^2}\right)\left(\frac{\partial w_h}{\partial x} - \Pi_K \frac{\partial w_h}{\partial x}\right) n_x\, d\gamma,$$

where $\Pi_K \frac{\partial w_h}{\partial x}$ denotes those linear functions defined along the vertical sides of the element K which take on the same value as the function $\frac{\partial w_h}{\partial x}$ at each vertex. Then we remark that inequality (3.9) would follow from "local" inequalities such as

(3.16) $$\left|E_{h,K}^{1,x}\left(u,\frac{\partial w_h}{\partial x}\right)\right| \leq C\, h\, |u|_{3,K}\, |w_h|_{2,K} \text{ for all } K\in\mathscr{E}_h,\ u\in H^3(K),\ w_h\in V_h.$$

But now inequality (3.16) irresistibly evokes an analogy with the Bramble-Hilbert lemma for linear forms, and indeed we have the following :

Lemma. Let Ω be an open bounded subset of R^n with a sufficiently smooth boundary, let k and ℓ be two integers, and let W be a space of functions satisfying the inclusions $P_\ell \subset W \subset H^{\ell+1}(\Omega)$; the space W is considered as being equipped with the norm $\|\cdot\|_{\ell+1,\Omega}$. Finally, let $A : H^{k+1}(\Omega)\times W \to R$ be a continuous bilinear form which satisfies

(3.17) $$A(u,v) = 0 \text{ for all } u \in P_k,\ v \in W,$$

(3.18) $$A(u,v) = 0 \text{ for all } u \in H^{k+1}(\Omega),\ v \in P_\ell.$$

Then there exists a constant $C = C(\Omega)$ such that

(3.19) $$|A(u,v)| \leq C \| A \| \, |u|_{k+1,\Omega} \, |v|_{\ell+1,\Omega} \text{ for all } u \in H^{k+1}(\Omega),\ v \in W.$$

Proof. For a given $v \in W$, the linear form $f_v : u \in H^{k+1}(\Omega) \to f_v(u) = A(u,v)$ is continuous and vanishes on the space P_k by (3.17). Thus, by the Bramble-Hilbert lemma [6], there exists a constant $C_1 = C_1(\Omega)$ such that

$$|f_v(u)| \leq C_1 \| f_v \|^*_{k+1,\Omega} \, |u|_{k+1,\Omega} \text{ for all } u \in H^{k+1}(\Omega),$$

$\| \cdot \|^*_{k+1,\Omega}$ being the dual norm of the norm $\| \cdot \|_{k+1,\Omega}$. To estimate $\| f_v \|^*_{k+1,\Omega}$, we use the characterization

$$\| f_v \|^*_{k+1,\Omega} = \sup_{u \in H^{k+1}(\Omega)} \frac{|A(u,v)|}{\| u \|_{k+1,\Omega}}.$$

By (3.18), we know that

$$|A(u,v)| = |A(u,v-p)| \text{ for all } p \in P_\ell$$

and thus

$$|A(u,v)| \leq \| A \| \, \| u \|_{k+1,\Omega} \inf_{p \in P_\ell} \| v-p \|_{\ell+1,\Omega}.$$

The proof of the lemma is then achieved by using the equivalence of the quotient norm $\inf_{p \in P_\ell} \| v-p \|_{\ell+1,\Omega}$ and of the semi-norm $|v|_{\ell+1,\Omega}$ over the quotient space $H^{\ell+1}(\Omega)/P_\ell$ (cf. Nečas [18]).

In order to apply the above Lemma, we use the fact that

(3.20) $$\oint_{\partial K} \left(\frac{\partial w_h}{\partial x} - \Pi_K \frac{\partial w_h}{\partial x}\right) n_x \, d\gamma = 0 \text{ for all } K \in \mathcal{E}_h,\ w_h \in V_h,$$

as a simple computation shows. From (3.20) and (3.15) respectively, we deduce that

(3.21) $$E^{1,x}_{h,K}\left(u, \frac{\partial w_h}{\partial x}\right) = 0 \text{ for all } u \in P_2,\ \frac{\partial w_h}{\partial x} \in W_K,$$

(3.22) $$E^{1,x}_{h,K}\left(u, \frac{\partial w_h}{\partial x}\right) = 0 \text{ for all } u \in H^3(K),\ \frac{\partial w_h}{\partial x} \in P_0,$$

where $W_K = \{\frac{\partial w_h}{\partial x} : K \to R : w_h \in V_h\}$. Using changes of variables from the finite element K to a reference finite element (over which the Lemma is applied) as in Lemma 3 of Crouzeix & Raviart [11], we obtain inequality (3.16).

If we add up relations (3.21) and analogous relations "in y", we see that

(3.23) $$E_h(u,w_h) = 0 \text{ for all } u \in P_2,\ w_h \in V_h,$$

which is precisely the patch test in its "global" form, the relations (3.21) being a "local" form thereof.

In table 2, three types of nonconforming finite elements are given, along with

TABLE 2

Some nonconforming finite elements for the plate problem

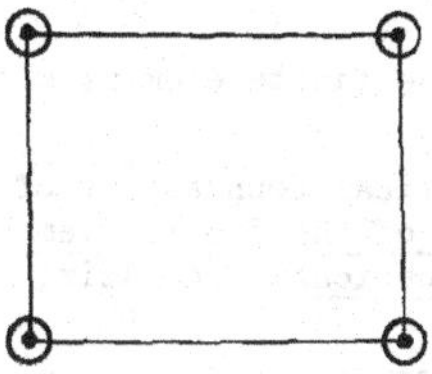 "Adini's rectangle"	$P_K = \{p:(x,y) \to p(x,y) = \sum_{0\leq i+j\leq 3} \alpha_{ij}x^iy^j + \alpha_{13}xy^3 + \alpha_{31}x^3y\}$. dim $P_K = 12$; $P_3 \subset P_K$. $\Vert u-u_h \Vert_h = O(h)$ if $u \in H^3(\Omega)$. Origin : Adini & Clough [1]. Error bounds : Lascaux & Lesaint (to appear).
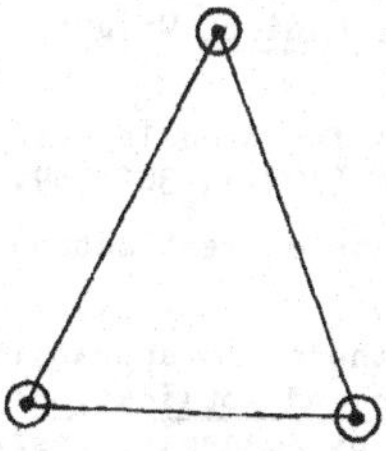 "Zienkiewicz' triangle"	$P_K = \{p \in P_3;$ internal node eliminated$\}$. dim $P_K = 9$; $P_2 \subset P_K$. The patch test is passed if all sides of all triangles are parallel to three directions only. $\Vert u-u_h \Vert_h = O(h)$ if $u \in H^3(\Omega)$. Origin : Bazeley, Cheung, Irons & Zienkiewicz [3]. Error Bounds : Lascaux & Lesaint (to appear).
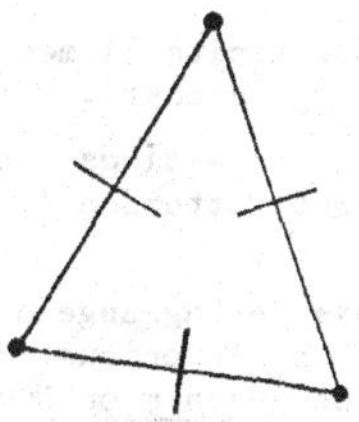	$P_K = P_2$. dim $P_K = 6$. The patch test is always passed. $\Vert u-u_h \Vert_h = O(h)$ if $u \in H^4(\Omega)$. Origin : Morley [17]. Error bounds : Lascaux & Lesaint (to appear).

conditions insuring that the patch test is passed. Let us also mention for completeness that other types of nonconforming methods have been considered by various authors: penalty methods (cf. Babuska & Zlámal [3]), hybrid methods (cf. Pian [19] and a forthcoming paper of Brezzi), decomposition methods (cf. Glowinski [13] and Ciarlet & Raviart (to appear)), mixed methods (cf. Johnson [15]).

REFERENCES

[1] Adini, A.; Clough, R.W. : Analysis of plate bending by the finite element method NSF Report G. 7337, 1961.

[2] Babuška, I.; Aziz, A.K. : Survey Lectures on the Mathematical Foundations of the Finite Element Method, The Mathematical Foundations of the Finite Element Method with Applications to Partial Differential Equations (A.K. Aziz, Editor), pp. 3-359, Academic Press, New York, 1972.

[3] Babuška, I.; Zlámal, M. : Nonconforming elements in the finite element method, Technical Note BN-729, University of Maryland, College Park, 1972.

[4] Bazeley, G.P.; Cheung, Y.K; Irons, B.M.; Zienkiewicz, O.C. : Triangular elements in bending-conforming and nonconforming solutions, Proceedings Conference on Matrix Methods in Structural Mechanics, Wright Patterson A.F.B., Ohio, 1965.

[5] Bogner, F.K.; Fox, R.L.; Schmit, L.A. : The generation of interelement compatible stiffness and mass matrices by the use of interpolation formulas, Proceedings Conference on Matrix Methods in Structural Mechanics, Wright Patterson A.F.B., Ohio, 1965.

[6] Bramble, J.H.; Hilbert, S.H. : Bounds for a class of linear functionals with applications to Hermite interpolation, Numer. Math. 16 (1971), 362-369.

[7] Bramble, J.H.; Zlámal, M. : Triangular elements in the finite element method, Math. Comp. 24 (1970), 809-820.

[8] Ciarlet, P.G. : Orders of convergence in finite element methods. To appear in Proc. Conference on the Mathematics of Finite Elements and Applications; Brunel University, April 18-20, 1972, to be published by Academic Press.

[9] Ciarlet, P.G.; Raviart, P.-A. : General Lagrange and Hermite interpolation in R^n with applications to finite element methods, Arch. Rational Mech. Anal. 46 (1972), 177-199.

[10] Clough, R.W.; Tocher, J.L. : Finite element stiffness matrices for analysis of plates in bending, Proceedings Conference on Matrix Methods in Structural Mechanics, Wright Patterson A.F.B., Ohio, 1965.

[11] Crouzeix, M.; Raviart, P.-A. : Conforming and nonconforming finite element methods for solving the stationary Stokes equations. I (to appear).

[12] Fraeijs de Veubeke, B. : Bending and stretching of plates, Proceedings Conference on Matrix Methods in Structural Mechanics, Wright Patterson A.F.B., Ohio, 1965.

[13] Glowinski, R. : Approximations externes, par éléments finis de Lagrange d'ordre un et deux, du problème de Dirichlet pour l'opérateur biharmonique. Méthode itérative de résolution des problèmes approchés, Conference on Numerical Analysis, Royal Irish Academy, 1972.

[14] Irons, B.M. ; Razzaque, A. : Experience with the patch test for convergence of finite elements, The Mathematical Foundations of the Finite Element Method with Applications to Partial Differential Equations (A.K. Aziz, Editor), pp. 557-587, Academic Press, New York, 1972.

[15] Johnson, C. : On the convergence of some mixed finite element methods in plate bending problems (private communication).

[16] Landau, L.; Lifchitz, E. : Théorie de l'Elasticité, Mir, Moscou, 1967.

[17] Morley, L.S.D. : The triangular equilibrium element in the solution of plate bending problems, Aero. Quart. 19 (1968), 149-169.

[18] Nečas, J. : Les Méthodes Directes en Théorie des Equations Elliptiques, Masson, Paris, 1967.

[19] Pian, T.H.H. : Finite element formulation by variational principles with relaxed continuity requirements, The Mathematical Foundations of the Finite Element Method with Applications to Partial Differential Equations (A.K. Aziz, Editor), pp. 671-687, Academic Press, New York, 1972.

[20] Raviart, P.-A. : Méthode des Eléments Finis (Lecture Notes), Université de Paris VI, Paris, 1972.

[21] Sander, C. : Bornes supérieures et inférieures dans l'analyse matricielle des plaques en flexion-torsion, Bull. Soc. Roy. Sci. Liège 33 (1964), 456-494.

[22] Strang, G. : Approximation in the finite element method, Numer. Math. 19 (1972), 81-98.

[23] Strang, G. : Variational crimes in the finite element method, The Mathematical Foundations of the Finite Element Method with Applications to Partial Differential Equations (A.K. Aziz, Editor), pp. 689-710, Academic Press, New York, 1972.

[24] Strang, G.; Fix, G. : An Analysis of the Finite Element Method, Prentice-Hall, Englewood Cliffs, 1973.

[25] Zienkiewicz, O.C. : The Finite Element Method in Engineering Science, McGraw-Hill, London, 1971.

[26] Zlámal, M. : On the finite element method, Numer. Math. 12 (1968), 394-409.

Discretization and chained approximation

L. Collatz

Summary: Different kinds of approximation, especially chained approximation, are illustrated on linear and nonlinear differential equations. Particularly possibilities of getting exact error bounds are considered.

Let us compare discretization methods and parametric methods.

1. Discretization

(Difference methods, finite element methods and others - compare Mitchell [69].)

For complicated problems, for instance with systems of nonlinear partial differential equations, usually only finite difference methods and finite element methods will be practicable. But in these cases it is possible often to spare much computer time

a) by using better approximation formulas (for instance "Mehrstellenverfahren" for boundary and for initial value problems, compare Collatz [60], [72] etc.),

b) by using different meshsizes, compare Collatz [65], Whiteman-Gregory [73],

c) for problems which are really three-dimensional one could use parametric methods or mixed parametric-discrete methods.

The finite element method, compare Zienkiewicz [71], Whiteman [73], Whiteman-Gregory [73] etc., is treated in many lectures of this meeting and will therefore not be considered more in this paper.

2. Parametric methods

The principle of monotonicity (if it is applicable) gives a onesided Tschebyscheff Approximation (T.A.) problem (compare Collatz [70], Krabs [72] etc.,) and has great advantages compared with the maximum principle. The monotonicity gives in many cases better bounds and is applicable often in nonlinear cases.

Example 1: (Ideal flow of a liquid between walls, fig.1)

We consider the boundary value problem for a function $u(x,y)$

$$\Delta u = \frac{\partial^2 u}{\partial x^2} + \frac{\partial^2 u}{\partial y^2} = 0 \quad \text{in } B \quad (|x| \leq 2,\ 0 < y < \psi(x) = 1 + \frac{x}{4} - \frac{x^3}{48})$$

and the boundary conditions:

$$u = 0 \quad \text{for} \quad y = 0; \qquad u = 1 \quad \text{for} \quad y = \psi(x)$$

$$u = \frac{3}{2}y \quad \text{for} \quad x = -2; \qquad u = \frac{3}{4}y \quad \text{for} \quad x = 2$$

We look for an approximate solution of the form

$$u \approx w = \sum_{\nu=1}^{p} a_\nu \phi_\nu(x,y) \quad \text{with given functions } \phi_\nu(x,y)$$

which satisfy the differential equations, for instance

$$\phi_1 = y, \quad \phi_2 = xy, \quad \phi_3 = y^3-3x^2y$$

$$\Phi_4 = y^3x-x^3y, \qquad \phi_5 = y^5-10x^2y^3+5x^4y^4$$

$$\phi_6 = 3xy^5-10x^3y^3+3yx^5.$$

The coefficients a_ν are determined in such a way that the modulus of the error $\varepsilon = w-u$ on the boundary ∂B is as small as possible (T.A.). Then the boundary maximum principle gives the exact error bounds:

$$|\varepsilon| \leq \delta \text{ on } \partial B \text{ has the consequence: } |\varepsilon| \leq \delta \text{ in } B.$$

For the numerical calculation the problem is written as a linear optimization problem (and then discretized):

$$-\delta \leq w(x,a_\nu) - u \leq \delta \quad \text{for } x \in \partial B, \quad \delta = \text{Min}$$

which is solved by using Simplex and related methods, compare Laurent [72]. I thank Mr. Günther and Miss Moldenhauer for the computation. The result is for $p = 6$

$$\begin{aligned} a_1 &= 1.02085161 & a_2 &= -0.24020364 \\ a_3 &= -0.01683085 & a_4 &= -0.01296090 \\ a_5 &= -0.00144703 & a_6 &= 0.00008421, \end{aligned}$$

the error bound is

$$|w - u| \leq 0.00538 \text{ in } B .$$

3. Nonlinear differential equations

Example 2: Distribution of the temperature $u(x,y)$ with respect to chemical reactions, for which there exists a critical temperature, say $u=1$, so that the produced heat may simply be given by $(1-u)^{-1}$. Suppose the boundary value problem as

$$\Delta u = \frac{1}{1-u} \text{ in } B \text{ for } r < 1 \text{ with } r^2 = x^2 + y^2 + z^2$$

$$u = 0 \text{ on } \partial B \text{ for } r = 1 .$$

Starting with a function $v_0(x,y,z)$ with $v_0 = 0$ on ∂B we determine a function v_1 with $\Delta v_1 = \frac{1}{1-v_0}$ in B, $v_1 = 0$ on ∂B.

If we choose $v_1 = \sum_{\nu=1}^{p} a_\nu \phi_\nu(x,y,z)$, we get an approximation problem

$$0 \leq v_1 - v_0 \leq \delta, \quad \delta = \text{Min}.$$

(The boundary maximum principle gives here no error bounds.)
We put

$$v_1 = a_1(1-r^2) + a_2(1-r^4)$$

and choose a_2 so that

$$v_0 = 0 \quad \text{for } r = 1:$$

$$v_0 = (1-r^2)\frac{6a_1 - 1}{r^2 + 6a_1(1-r^2)}$$

$$v_1 = \frac{(1-r^2)}{20}[1 + r^2 + a_1(14-6r^2)].$$

(Continuation in No.4.)

4. Chained approximation

An approximation problem is called a "chained approximation" if at least one of the parameters a_ν occurs at (at least) two places in the analytic expression for w (and if one cannot avoid this phenomenon by a one to one transformation of the parameters).

In example 2 we get for $v_1 - v_0$ an expression in which a_1 occurs at different places and therefore we have a chained approximation.

In this example we get the solution from the equation
$\frac{dv_o}{dr} = \frac{dv_1}{dr}$ for $r = 1$: $a_1 = \hat{a}_1 = \frac{11}{56}$.

By substituting w_o, w_1, b_1 instead of v_o, v_1, a_1 we get an upper bound for u ; the best value $b_1 = \hat{b}_1$ is
$\hat{b}_1 = \frac{1}{84}(57 - \sqrt{1569})$; it is $v_1(a_1=\hat{a}_1) \leq w_1(b_1=\hat{b}_1)$ satisfied and therefore exists at least one solution u with

$$v_1(a_1=\hat{a}_1) \leq u \leq w_1(b_1=\hat{b}_1),$$

particularly

$$\frac{3}{16} = 0.1875 \leq u(0,0) \leq 0.1950.$$

For this result Schauder's fixed point theorem was used, cf. Collatz [66], p.355. Better results can be got with more parameters.

5. Inverse problems in differential equations

Chained approximation occurs frequently.

Example 3: Inverse problem. For instance let us observe the density or concentration of temperature $u(x,t)$ at two times $t = 0$ and $t = 1$

$$u(x,0) = f(x), \quad u(x,1) = g(x)$$

and assume that $u(x,t)$ satisfies a differential equation of the form

$$\frac{\partial^2 u}{\partial x^2} - k\frac{\partial u}{\partial t} = 0$$

and that $u(x,t)$ is of the form

$$u(x,t) = \frac{1}{\sqrt{t+t_o}} \exp\left[-\frac{k}{4}\frac{(x-x_o)^2}{t+t_o}\right].$$

We consider k and t_o as unknowns we wish to determine, and f,g are given functions from observations. Again we have a chained approximation.

Special cases of nonlinear chained approximation were treated by Wachspress [68].

6. Sufficient conditions for a best chained approximation

With the aid of the theory of H-sets of the theory of approximation (Collatz-Krabs [73]) one can get the following result, we describe very briefly:

Let us write

$$v_1(x_1,\dots,x_n,a_1,\dots,a_p)-v_o(x_1,\dots,x_n,a_1,\dots,a_p)=\phi(x_j,a_\nu)$$

for $(x_1,\dots,x_n) \in B$.

Let us consider two sets a',a'' of parameters $a_1',\dots,a_p'$, $a_1'',\dots,a_p''$.

We call a set of points P_μ $(\mu=1,\dots,m)$, $Q_\varrho(\varrho=1,\dots,r)$ an H-set for the problem

$$0 \leq v_1 - v_o \leq \delta, \quad \delta = \text{Min} ,$$

if there exists no pair a',a'' of parameters with

$$\phi' - \phi'' \begin{cases} \leq 0 & \text{in all points } P_\mu \\ > 0 & \text{in all points } Q_\varrho . \end{cases}$$

If we have a certain parametervector $\hat{a} = (\hat{a}_1,\dots,\hat{a}_p)$ and $\hat{\phi} = \phi(x_j,\hat{a}_\nu)$ with

$$\hat{\phi}(P_\mu) = 0 \quad \text{for } \mu=1,\dots,m$$

$$\hat{\phi}(Q_\varrho) = \max_{x\in B} \hat{\phi} \quad \text{for } \varrho=1,\dots,r ,$$

$$0 = \min_{x\in B} \hat{\phi}$$

then $v_o(x_j,\hat{a})$, $v_1(x_j,\hat{a})$ is a best chained approximation.

7. Algorithms for linear approximation

In this case one gets the same algorithms as they are known for linear onesided Tschebyscheff-Approximation.

Example 4: For the problem for a function $u(x,y)$ with $r^2 = x^2+y^2 = s = \sigma - 1$

$$-\Delta u = (1+r^2)u \quad \text{in } B \ (r<1)$$
$$u = 1 \quad \text{on } \partial B \ (r=1) .$$

We are using functions v_o, v_1, w_1, w_o as in No.4;

$$-\Delta v_1 = (1+r^2)v_o \text{ in } B \quad , \quad v_1 = 1 \text{ on } \partial B$$
$$v_o = 1 + (4a_1-1)\frac{1-s}{1+s} , \quad v_1 = 1 + a_1(1-s) + \frac{1-2a_1}{8}(1-s^2)$$
$$0 \le \phi \le \delta, \ \delta = \text{Min} , \quad \phi = \frac{1-s}{\sigma}[a_1(\frac{-\sigma^2+4\sigma-16}{4}) + \frac{\sigma^2}{8} + 1].$$

We get the exact lower and upper bounds for the solution u, fig.2,

$$\frac{36-9s-s^2}{26} \le u \le \frac{3-s}{2}$$

for instance

$$\frac{18}{13} \approx 1.3846 \le u(0,0) \le 1.5.$$

As H-set we take the points $\sigma=1$ and a σ with $1<\sigma<2$ (compare the method Collatz [66], 420-430).

The condition $\sigma=1$ or $v_1-v_o \ge 0$ requires $a_1 \le \frac{9}{26}$,
the condition with $1<\sigma<2$ and $v_1-v_o<0$ requires $a_1 > \frac{9}{26}$;
this gives a contradiction and therefore v_o, v_1 is a best solution for $0 \le v_1-v_o \le \delta$, $\delta =$ Min.

In more complicated cases it may be necessary to approximate simultaneously in the interior of B and on the boundary ∂B, compare Bredendiek [70], Collatz-Krabs [73].

8. Eigenvalue problems

Example 5: For a function $u(r,\varphi)$ may be given

$$Lu = \frac{\partial}{\partial r}\left[(1+r^2)\frac{\partial u}{\partial r}\right] = -\lambda u \text{ in } B\ (r<1),\quad u+\frac{\partial u}{\partial r} = 0 \text{ on } \partial B\ (r=1).$$

We consider the first eigenvalue λ_1 and a positive eigenfunction u in B. We take as approximation

$$u \approx w = 1 + \sum_{\nu=1}^{p} a_\nu r^{2\nu},$$

and have the chained approximation

$$\left| \frac{1+\sum_{\nu=1}^{p} a_\nu r^{2\nu}}{\sum_{\nu=1}^{p} a_\nu L(r^{2\nu})} + \Lambda \right| \leq \delta, \quad \delta = \text{Min}$$

with the unknowns a_ν, Λ, δ.

9. Relative approximation

Example 6: Eigenvalue problem for a function $y(x)$:

$$-[f(x)\,y'(x)]' = \lambda\, g(x)\, y(x) \quad \text{for} \quad x\in J=(a,b),\ y(a)=y(b)=0.$$

Let $f(x)$, $g(x)$ be given functions, positive for $x \in J$.

We compare with an exact solvable eigenvalue problem

$$-[p(x)\, y'(x)]' = \tilde{\lambda}\, \frac{y(x)}{p(x)} \quad \text{for} \quad x \in J,\ y(a) = y(b) = 0$$

with the known eigenvalues

$$\tilde{\lambda}_n = \left(\frac{n\pi}{q}\right)^2, \qquad q = \int_a^b \frac{ds}{p(s)}\ .$$

(Here $p(x)$ is an arbitrarily given positive continuous function.)

We wish to determine a function $k(x) = [p(x)]^{-1}$ in such a way that $p(x) \leq f(x)$, but $p(x)$ as near to $f(x)$ as possible, and $k(x)$ near to $\gamma g(x)$, where γ is unknown. With given f, g we have the unusual optimization problem for a positive function $k(x)$ and a constant $\gamma > 0$:

$$k(x) \geq \frac{1}{f(x)}\,, \quad k(x) \geq \gamma g(x) \ \text{for}\ x \in J\ ,$$

and

$$Q = \frac{1}{\gamma}\left[\int_a^b k(s)ds\right]^2 = \text{Min.}$$

Then we have lower bounds for all eigenvalues

$$\lambda_n \geq \frac{n^2\pi^2}{Q}\ .$$

10. Relative approximation for a partial differential equation

<u>Example 7:</u> Eigenvalue problem for a function $u(x,y)$:

$$Lu = -\frac{\partial}{\partial x}\left(P\frac{\partial u}{\partial x}\right) - \frac{\partial}{\partial y}\left(P\frac{\partial u}{\partial y}\right) = \lambda Pu \ \text{in}\ B \quad (0<x<1,\ 0<y<1)$$

$$u = 0 \ \text{on}\ \partial B\ .$$

Here P may be $P = 1 + e^{x+y}$.

For $\hat{P} = c_1 \exp(c_2(x+y))$ instead of P, one could solve the eigenvalue problem exactly. Therefore we try to approximate P by $\hat{P}$ in the sense that the relative error is as small as possible, fig.3:

$$(1-\alpha)\sqrt{2}\, e^{bs} \leq \sqrt{1+e^{s}} \leq \sqrt{2}\, e^{bs} \quad \text{in } 0 \leq s \leq 2 \quad \text{with } s=x+y.$$

We get $\alpha = 0.06$, $b = 0.3585$, $2b^2 = 0.2570$ and the inclusion for infinitely many eigenvalues

$$(1-\alpha)^2 D \leq \lambda_{n,m} \leq \frac{1}{(1-\alpha)^2} D \quad \text{with}$$

$$D = 0.2570 + \pi^2(n^2+m^2) \quad (n,m=1,2,3,\ldots).$$

1. Further applications

There are many other problems of numerical analysis as applications of chained approximation; we mention only some fields:

singular boundary value problems, linear and nonlinear integral equations, nonlinear vibrations etc.

References

E. Bredendiek [70], Charakterisierung und Eindeutigkeit bei simultanen Approximationen, Z. angew. Math. Mech. 50(1970), 403-410.

L. Collatz[60], The numerical treatment of differential equations, Springer 1960, 568 p.

L. Collatz [65], Über einige neue Entwicklungen bei Differenzenverfahren für Randwertaufgaben partieller Differentialgleichungen, Aplikace Matematiky, Prag, 10(1965), 165-177.

L. Collatz [66], Functional Analysis and Numerical Mathematics, Acad. Press 1966, 473 p.

L. Collatz [70], Einseitige Tschebyscheff Approximation bei Randwertaufgaben, Proc. Internat. Conference on Constructive funct. Theory, Varna, Bulgaria, 1970, 151-162.

L. Collatz [72], Hermitean methods for initial value problems in partial differential equations, to appear Proc. Symp. Numer. Methods, Dublin, Eire, 1972.

L. Collatz [73], Chained Approximation, to appear Proc. Symp. Funct. Anal. and Applicat., Madras, India, 1973.

L. Collatz - W. Krabs [73], Tschebyscheff Approximation, Teubner, 1973, to appear.

W. Krabs [72], Bemerkungen zur einseitigen Tschebyscheff Approximation, Computing 9(1972), 335-342.

P. J. Laurent [72], Approximation et optimisation, Paris 1972, 531 p.

A. R. Mitchell [69], Computational methods in partial differential equations, John Wiley & Sons, 1969, 255 p.

E. L. Wachspress [68], The ADI-Minimax problem, Proc. IFIP Congress, Edinburgh, 1968.

J. R. Whiteman [73], The mathematics of finite Elements and Applications, Acad. Press, 1973, 520 p.

J. R. Whiteman - J. A. Gregory [73], Mesh Refinement in Finite Element Methods, to appear, Proc. Symp. Numer. Solut. Diff. Equ., Dundee, 1973.

O. C. Zienkiewicz [71], The Finite Element Method in Engineering Science, McGraw Hill, 1971, 521 p.

Fig. 1

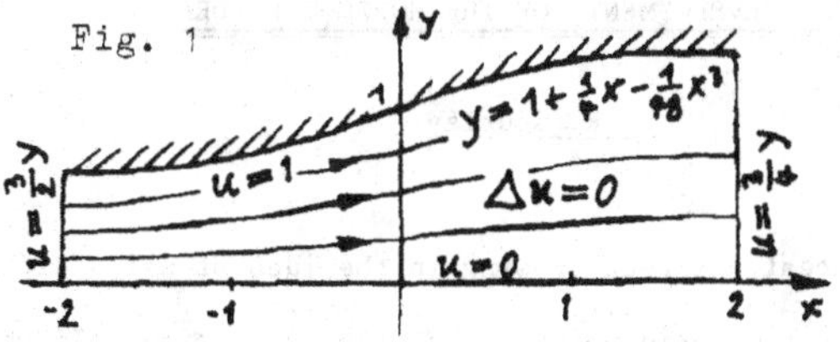

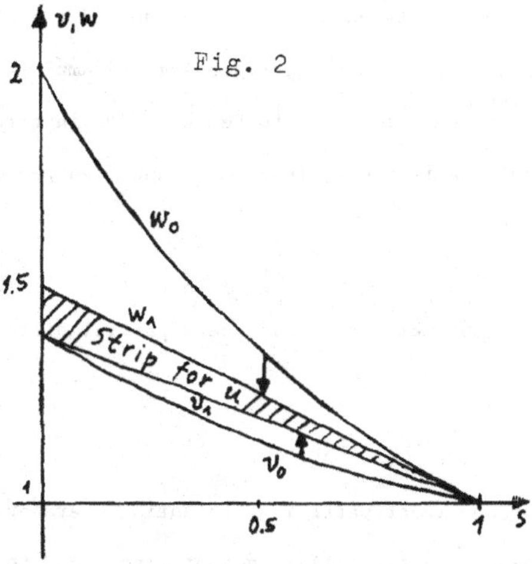

Fig. 2

Fig. 3

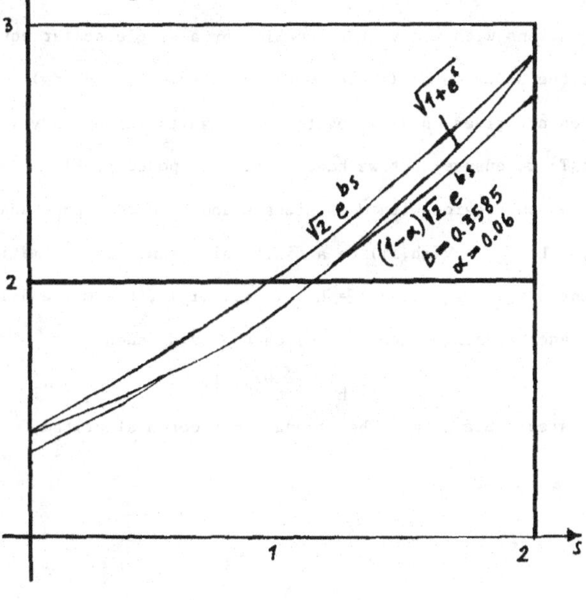

RECENT DEVELOPMENTS OF THE HOPSCOTCH IDEA

A. R. Gourlay

In recent years interest has been growing in the idea of using different finite difference schemes at different grid points in a regular fashion. In this paper a survey will be given of some of the recent theoretical work on this topic but the main emphasis will be on indicating the value of the freedom of the approach in designing a computational scheme for a particular problem. Examples will be given of situations where a hopscotch approach seems to be effective in circumventing awkward problem features such as anisotropy, interfaces and open regions.

1. Description of Hopscotch

To describe hopscotch we consider the evolutionary partial differential equation

$$\frac{\partial u}{\partial t} = F(u) \tag{1}$$

where $F(\bullet)$ may be a linear or nonlinear partial differential operator in a number n of space variables. The type of equation (1) is not of importance to us in describing the hopscotch approach. Likewise it does not matter whether (1) is a system of equations with u a vector function or a single scalar equation. We suppose that the solution of (1) is required in some region $R(x) \times \{o \leq t \leq T\}$ and that initial values are given on $R(x)$ together with suitable boundary values on $\partial R(x) \times \{o \leq t \leq T\}$ to ensure that we have a properly posed problem. We superimpose a uniform grid on our region with mesh sizes h and τ in the space and time directions respectively. Let $V_i^m = V(ih, m\tau)$ be a finite difference approximation to $u(x,t)$ at the grid point $x = (x_1, x_2, \ldots, x_n) = h(i_1, i_2, \ldots, i_n)$, $t = m\tau$, and let $F_h(u)$ be a finite difference approximation to $F(u)$ on the grid, such that

$$|| F_h(u) - F(u) || \to 0$$

as $h \to 0$ in a suitable norm. The standard hopscotch algorithm for (1) is given by the finite difference scheme

$$V_i^{m+1} - \tau\theta_i^{m+1} F_h(V_i^{m+1}) = V_i^m + \tau\theta_i^m F_h(V_i^m) \qquad (2)$$

where the hopscotch mesh function θ_i^m is a zero-one variable whose value depends on the indices m and i, but which satisfies the conditions

$$\theta_i^{m+1} + \theta_i^m = 1 \qquad (3)$$

$$\theta_i^m \cdot \theta_i^{m+1} = 0 \quad .$$

Several choices of the θ function have already been investigated. In particular we note the following:-

(i) $\theta_i^m = m \pmod 2$,

a choice which reduces (2) to the Crank-Nicolson scheme on a 2τ grid, where the V_i^{m+1} values are intermediate to the calculation;

(ii) $\theta_i^m = (m+|i|) \pmod 2$

$|i| = i_1 + i_2 + \ldots + i_n$,

giving the odd-even hopscotch scheme discussed in Gourlay (1970), and equivalent to a multistage implementation of a Du Fort-Frankel scheme;

(iii) $\theta_i^m = (m + |i| - i_s) \pmod 2$

$s \in \{1,2,\ldots,n\}$

corresponding to the line-hopscotch method discussed in Gourlay and McGuire (1971).

The definition (iii) is probably the most important choice of θ function. The importance in the freedom of choice of the θ function is two-fold. Firstly, and we shall return to this point later, it allows the tailoring of the algorithm to the problem. Secondly the degree of implicitness in obtaining V_i^{m+1} from (2) is governed by the dependence of the θ variable on the index i. For example if we consider a linear second order parabolic problem in two space variables the choice (i) is well known to be heavily implicit, the choice (ii) can be organised to be explicit and the choice (iii) requires the solution of tridiagonal systems along half of the mesh lines parallel to one of the space axes. (Thus (iii) requires at each step only ½ of the work required in each step of the Peaceman Rachford algorithm.) For a nonlinear second order parabolic problem in two space variables (ii) requires the solution of nonlinear relations pointwise at ½ of the mesh points and (iii) leads to three term nonlinear systems again along ½ of the mesh lines. The accuracy of the algorithm tends to be increased as the implicitness is increased.

Although the choice (iii) is probably one of the best available one can extend the concept of hopscotch by introducing more than one θ function, and by using any decomposition properties of the F(u) function. For example for the two space dimensional heat equation one can decompose F(u) into the sum of an x_1 part and an x_2 part (just as in ADI methods). Then a decomposed hopscotch process can be defined as

$$V_i^{m+1} - \tau\theta_i^{m+1} F_h^{(1)}(V_i^{m+1}) - \tau\zeta_i^{m+1} F_h^{(2)}(V_i^{m+1}) = V_i^m + \tau\theta_i^m F_h^{(1)}(V_i^m) + \tau\zeta_i^m F_h^{(2)}(V_i^m) \tag{4}$$

where θ_i^m, ζ_i^m both obey the equations (3). This process obviously includes the three choices (i), (ii), (iii) considered above by identifying $\zeta_i^m \equiv \theta_i^m$. Two choices of the mesh functions not satisfying this identity are of interest:-

(iv) $\theta_i^m = \zeta_i^{m+1} = m \pmod 2$

which gives the well known Peaceman-Rachford method with a time step of 2τ,

(v) $\theta_i^m = \zeta_i^{m+1} = (m+i_1) \pmod 2$

which gives the ADI hopscotch method.

The choice (v) produces a scheme with a degree of implicitness equal to that of the choice (iii), but which appears to have an accuracy lying between that of (iii) and (iv).

The stability and convergence of the methods (i) → (v) have been analysed in Gourlay and McGuire (1971) for second order parabolic problems. Hopscotch schemes for first order hyperbolic systems have been discussed in Gourlay and Morris (1972).

2. LOD Hopscotch

Since we can identify hopscotch with alternating direction methods, provided we interpret the meaning of direction in a very general sense, it is not surprising that we can develop methods of locally one dimensional (LOD) type.

Thus corresponding to the general hopscotch scheme (4) we have the LOD analogue given by

$$\begin{aligned} W_i^{m+1} - \tau\theta_i^m F_h^{(1)}(W_i^{m+1}) - \tau\zeta_i^m F_h^{(2)}(W_i^{m+1}) \\ = W_i^m + \tau\theta_i^m F_h^{(1)}(W_i^m) + \tau\zeta_i^m F_h^{(2)}(W_i^m) \quad . \end{aligned} \tag{5}$$

For the case of a linear problem where

$$F_h^{(1)}(W_i^m) = L_h^{(1)} . W_i^m , \; F_h^{(2)}(W_i^m) = L_h^{(2)} . W_i^m ,$$

we can show that the solutions of (4) and (5) are related by the equation

$$\left[I - \tau\theta_i^m L_h^{(1)} - \tau\zeta_i^m L_h^{(2)} \right] V_i^m = W_i^m \quad .$$

This relationship follows easily from the analysis in the paper of Gourlay and Mitchell (1972). Some of the above LOD schemes have computational advantages over their ADI counterparts. In addition those algorithms corresponding to the choices (ii) and (iii) above have no difficulty in picking up appropriate boundary data. This is usually a major problem when employing an LOD method. The stability, convergence and possible usefulness of LOD hopscotch schemes is discussed in the manuscript by Gourlay and Morris (1973).

3. Hopscotch for the Wave Equation

Recently Orley and McKee (1973) have considered the possibility of using hopscotch difference schemes for problems of the form

$$\frac{\partial^2 u}{\partial t^2} = L\,u. \tag{6}$$

[In fact their analysis only considered the case of $L \equiv -\frac{\partial^4}{\partial x^4}$.]

Their approach was basically to construct a three level scheme using an ADI hopscotch strategy. None of their schemes turned out to be an improvement on the simple explicit scheme. [Some of their schemes were in fact generalisations of hopscotch as defined above, in that the θ variable was chosen accordingly to a rule like

$$\theta_i^m = \begin{cases} 1 & \text{if } m+i = \rho \pmod 3 \\ 0 & \text{otherwise} \end{cases}$$

for $\rho \in \{0,1,2\}$.]

An alternative approach which appears to offer more hope in deriving hopscotch schemes with good stability properties is based on reducing (6) to a system of the form (1). Thus we write $u_t \equiv \frac{\partial u}{\partial t}$ and solve (6) as the system

$$\frac{\partial \underset{\sim}{u}}{\partial t} = \frac{\partial}{\partial t}\begin{pmatrix} u \\ u_t \end{pmatrix} = \begin{pmatrix} 0 & 1 \\ L & 0 \end{pmatrix}\begin{pmatrix} u \\ u_t \end{pmatrix} = A\underset{\sim}{u}\,. \tag{7}$$

Hopscotch applied to (7) then gives the scheme

$$[I - \tau\theta_i^{m+1} A_h]\underset{\sim}{V}_i^{m+1} = [I + \tau\theta_i^m A_h]\underset{\sim}{V}_i^m$$

where $\underset{\sim}{V}_i^m = (V_i^m, (V_t)_i^m)^T$ is an approximation to the vector $(U,U_t)^T$ at the point (i,m). In terms of components the above equation may be written as

$$V_i^{m+1} - V_i^m = \tau\left(\theta_i^{m+1}(V_t)_i^{m+1} + \theta_i^m (V_t)_i^m\right)$$

$$(V_t)_i^{m+1} - (V_t)_i^m = \tau(\theta_i^{m+1} L_h V_i^{m+1} + \theta_i^m L_h V_i^m).$$

Careful use of these two equations enables the quantities $(V_t)_i^{m+1}$, $(V_t)_i^m$ to be eliminated, and the resulting difference scheme for V, and hence the hopscotch difference replacement of (7) is found to be

$$\left[1 - \tau^2\theta_i^m L_h\right]V_i^{m+2} - 2V_i^{m+1} + \left[1 - \tau^2\theta_i^m L_h\right]V_i^m = 0 \qquad (8)$$

which of course is an LOD hopscotch scheme. [It is interesting to note that if we care to look at pages 216-218 of the book by Mitchell (1969) we see that the above equation corresponds to an LOD like scheme whereas the approach of Orley and McKee corresponds to a locally alternating direction scheme.] For the wave equation, (8) appears to be unconditionally stable and explicit, and convergent for $\tau, h \to 0$ when the ratio τ/h^2 is held constant.

For the equation of vibration where L is fourth order it is quite possible that (8) will be unconditionally stable. However for such an equation it will require the solution of tridiagonal systems at each step. The above discussion covers work which is at a very preliminary stage. However we now appear to understand better how to apply the hopscotch approach to problems with time derivatives higher than the first order.

4. Applications

In the second part of this paper we will attempt to show how a hopscotch approach can be valuable in designing an algorithm to suit a particular type of problem.

4.1 Anisotropic problems

Firstly we will report on some work carried out at Dundee by Morris and Nicoll (1973) on the use of hopscotch in solving heat conduction problems in anisotropic media. The problems they considered were quite complex so we will simplify matters to ease discussion.

We consider solving the three dimensional heat equation

$$\frac{\partial u}{\partial t} = \kappa_1 \frac{\partial^2 u}{\partial x^2} + \kappa_2 \frac{\partial^2 u}{\partial y^2} + \kappa_3 \frac{\partial^2 u}{\partial z^2} + g(x,y,z,t)$$

where the coefficients κ_1, κ_3 are of the same magnitude but κ_2 is very much larger. [In the work of Morris and Nicoll the ratio of κ_2 to κ_1 was typically of the order of 100.] This problem is therefore highly anisotropic in the y direction. In the description of the line hopscotch method, or the ADI hopscotch method, the user is given the freedom of choice of direction of implicitness. In this case one can choose any of the directions x,y,z as the direction of implicitness. Of course the local error of the hopscotch scheme is dependent on the particular choice. Morris and Nicoll found experimentally and verified theoretically that much more accurate results could be obtained by aligning the direction of implicitness with the y axis. Typical of the results they quote is an error of $O(10^{-3})$ with implicitness in the y direction against an error of $O(10^{-1})$ with implicitness otherwise. They also found that the choice (ii) gave inaccurate results. It might also be conjectured that a standard ADI procedure would also be inaccurate for anisotropic problems. Morris and Nicoll also apply the various hopscotch techniques to solving a thermal print head problem, with considerable success.

4.2 Open Regions

In this section our aim is to show that a hopscotch algorithm can be used in a neat way to solve problems on an open region in the space variables. To be specific let us suppose we require to solve a time dependent problem in two space variables over the region $0 \leq x \leq 1$, $y \geq 0$. In our discussion we will concentrate on the use of line hopscotch. Odd-even hopscotch (or the DuFort-Frankel method) generally has inferior accuracy to line hopscotch and for this reason we prefer to use a line hopscotch approach. For this problem it is natural to use implicitness parallel to the x-axis and compute outwards in direction of increasing y, until we reach a region of "negligible change". Our θ contours in this example are thus the mesh lines y = constant as shown in Fig. 1. In the figures we indicate contours with the same θ value by the same symbol. Thus we only require to use two symbols X and

O. It is immaterial which of these corresponds to the value of $\theta = 1$ and which to the value $\theta = 0$. Note that at the next time step the roles of the two sets of contours changes. That set of contours, say X, which corresponds to an explicit method at one level, will correspond to the implicit method at the following level. However if we have an interface at say $y = a$ together with some interface conditions then the above approach is not viable and we must now use a line hopscotch with the θ contours as shown in Figs. 2, 3, 4 depending on the grid. All these approaches lead to systems which are tridiagonal.

For problems over regions such as $x,y \geq 0$ the θ contours can be designed as indicated in Fig. 5 and again only tridiagonal systems require to be solved.

The stability of all these algorithms is covered by the theorems in Gourlay and McGuire (1971).

These simple examples help to show the usefulness of the hopscotch idea.

References

Gourlay, A. R. 1970 J. Inst. Maths. Applics. 6 375-390

Gourlay, A. R. and McGuire, G. R. 1971 J. Inst. Maths. Applics. 7 216-227

Gourlay, A. R. and Mitchell, A. R. 1972 SINUM 6 37-46

Gourlay, A. R. and Morris, J. Ll. 1972 IBM J. Res. Dev. 16 349-353

Gourlay, A. R. and Morris, J. Ll. 1973 Manuscript

Mitchell, A. R. 1969 Computational Methods in Partial Differential Equations, Wiley.

Morris, J. Ll. and Nicoll, I. F. 1973 J. Comp. Phys. (to appear)

Orley, D. G. and McKee, S. 1973 J. Inst. Maths. Applics. 11 335-338

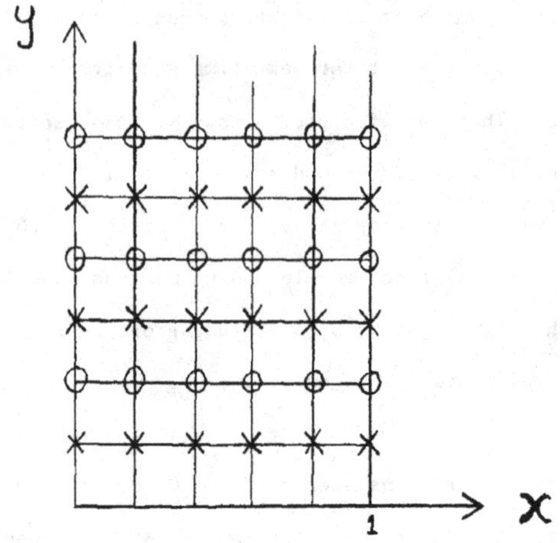

Figure 1

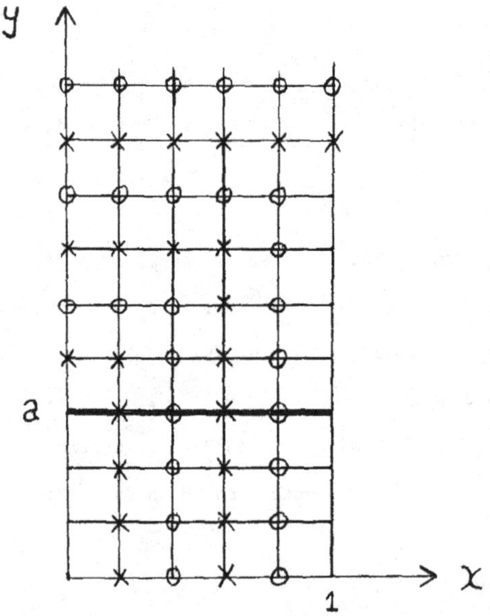

Figure 2

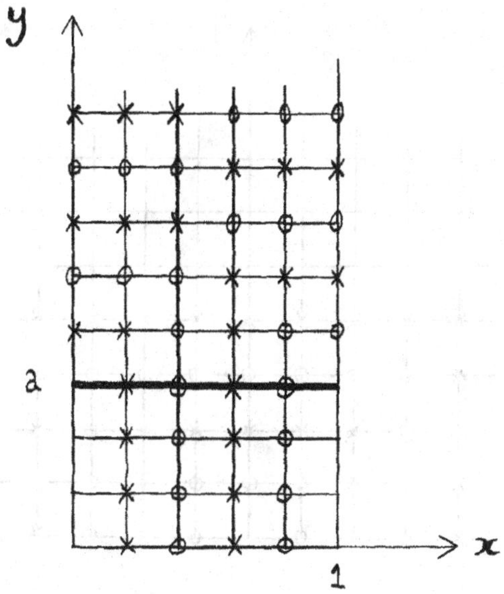

Figure 3

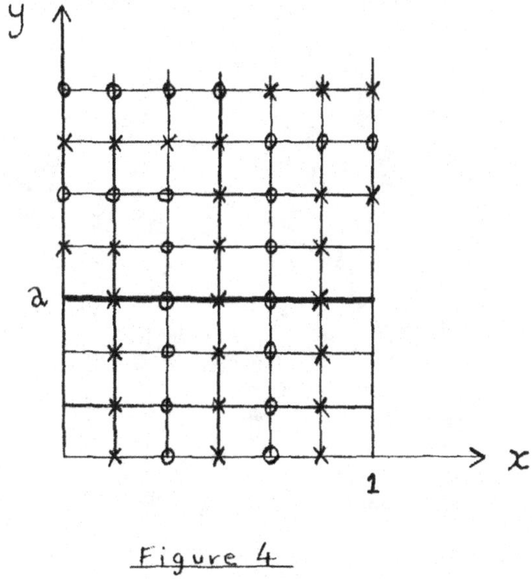

Figure 4

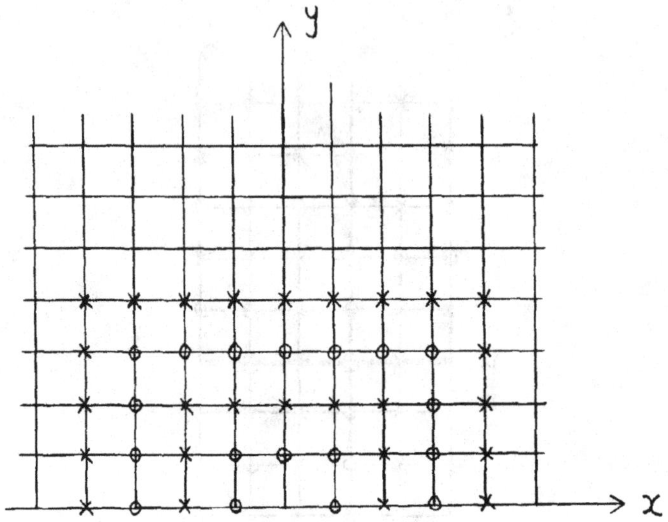

Figure 5

THE DEVELOPMENT OF SOFTWARE FOR SOLVING ORDINARY DIFFERENTIAL EQUATIONS

T.E. Hull

Abstract Factors involved in the development of good software are discussed, with particular reference to programs for solving ODE's. These factors include the basic structuring of the programs themselves, along with the appropriateness of various language facilities, comparisons of efficiency, proofs of correctness, certification and distribution, etc.

Introduction

There are several attributes one would like to have in good software packages. For one thing, programs should be convenient to use and easy to understand. They should also be reliable, efficient, and provably correct.

The purpose of this paper is to indicate the extent to which several of us at the University of Toronto believe we have been able to reach these goals in the development of programs for solving ordinary differential equations.

The discussion will be limited to programs for systems of first order equations with initial conditions, and it will be assumed that the equations present no special difficulties, such as those associated with stiffness or discontinuities. Moreover, it is assumed that the programs are intended to meet only the basic requirements of a user, so that provision need not be made for special features, such as monitoring intermediate solution values. We have also been considering these other possibilities, especially stiffness, but our results in these areas are still incomplete.

We have recently completed some programs based on Runge-Kutta-Fehlberg formulas, and we are redesigning variable-order-Adams and extrapolation programs, earlier versions of which have been tested quite carefully. I believe that these programs meet, to a considerable degree, the five goals of convenience, understandability, reliability, efficiency and correctness.

Although the following sections deal with these attributes in relation to programs for ordinary differential equations, the conclusions are applicable, at least to some extent, in other areas as well.

Structure of programs

The key to much that is desirable in a good program is its structure. If it is well-structured, it is easy to understand (and to modify if necessary), and it is relatively easy to prove correct.

We have developed a common structure for all our programs for ordinary differential equations. Initial ideas about this structure were described in a paper presented to the IFIP meeting in Edinburgh [5]. These were modified and made more specific for a paper on comparing methods [7]. The present form will be described in detail, along with at least one program listing, in a paper now being completed [9].

To discuss the main ideas in the structure we have adopted, we first need to specify the calling sequence, which is of the following form:

N, FCN, X, Y, XEND, TOL, IND, SCALE,
+ workspace + "own" variables

where the notation is mostly self-explanatory. The purpose of the program is to replace the initial value of X with XEND, and the initial value of Y with an approximation to Y at XEND, keeping the error per unit step less than the tolerance, TOL.

IND is an indicator that is set equal to 1 on initial entry. It is then changed by the program to permit efficient reentry with different values of XEND. (It is also used to signal error exits.)

We believe the user should also be expected to provide a measure of the scale of his problem. In [7] we proposed HMAX, the maximum allowed step-size. But this measure depends partly on the method, and it would be better if the user did not have to know so much about the method, especially if it is a variable order method. SCALE is used instead. It can be set quite arbitrarily for most problems, equal to 1.0, for example, but other settings may improve accuracy or efficiency. Larger values of SCALE will cause the program to be more cautious and may improve accuracy. The program uses SCALE to determine HMAX, and for best results, SCALE should be roughly equal to the appropriate Lipschitz constant for the problem.

Workspace will normally be declared by the user, if the program is to handle systems of variable size in an efficient way. Also "own" variables, such as the current step-size, and the error estimate, which would be useful on reentry to the program, may have to be declared by the user. (This is true, for example, in strict ANSI Fortran.)

The arguments N, FCN, ..., TOL are needed to define the problem. IND and SCALE are not absolutely essential, but they serve useful purposes, and SCALE moreover gives the user a simple measure of control over the method.

We can now outline the overall structure that we have adopted. It is shown in Figure 1.

```
 INITIALIZATION           (calculate HMIN, HMAX from TOL, SCALE, etc.)

...VALIDITY CHECK         (e.g., check HMIN and HMAX)

 REPEAT FOLLOWING UNTIL EXIT IN 4TH STAGE:

    PREPARE               (calculate step-size, also slope if necessary,
                               perhaps order)

    CALCULATE             (make calculations for one step)

    ESTIMATE              (estimate error)

.....DECIDE               (accept, then update and return if finished, or
                               error exit; otherwise try again)

 END OF REPEAT BLOCK
```

Figure 1. Overall structure of a program for ordinary differential equations. The dots indicate possible exits from the program.

The main point about Figure 1 is that the overall organization of the program is relatively easy to understand. Moreover, since each part of the program shown in Figure 1 is relatively self-contained, logically, further refinements of the program will also be easy to understand. For example, all calculations related to the choice of step-size are collected together in one place. This kind of organization makes it much easier to prove a program's correctness.

Language facilities

To implement a program we must of course express it in a particular programming language. In attempting to develop well-structured programs we have become increasingly aware of the relative merits and limitations of various programming languages.

We have needed only four basic constructs, plus the ability to exit, from anywhere within a particular construct to whatever outer level of nesting we wish. The constructs are of the form IF-THEN-ELSE-END, DO-END, REPEAT-END, and CASE-END. For example, the CASE statement is used in the PREPARATION stage of our programs; in the simplest program three cases have to be considered when calculating the trial step-size: on initial entry, after a successful step, and after a failure. IND is used to indicate which case is to be considered.

With these constructs, and the exit facility, we do not need any labels or any GOTO statements. However, we are forced to use existing languages (mostly Fortran in our case). We have therefore decided to use these languages only in a very disciplined way, so that the basic constructs are implemented only in a very restricted way. Our conventions in using Fortran are so restrictive that, if you know the simple and rather obvious rules that have been imposed, it is possible to understand our programs even if all the statement numbers and GOTO statements have been removed. In fact, it is easier if they have been removed!

To illustrate, Figure 2 shows how we implement the IF-THEN-ELSE-END construct in Fortran. Beside it is shown what is left if all statement numbers and GOTO statements are removed. The C's have also been removed, and CONTINUE has been replaced by END. (We actually write many of our programs in the form shown on the right; a

```
      IF(boolean) GO TO n1
         GO TO n2
C     THEN
 n1      _____
         _____
         _____
         GO TO n3
C     ELSE
 n2      _____
         _____
         _____
 n3   CONTINUE
```

```
IF(boolean)
THEN
   _____
   _____
   _____
ELSE
   _____
   _____
   _____
END
```

Figure 2. An implementation of IF-THEN-ELSE-END in Fortran, along with the corresponding DEFT version.

processor called DEFT [12] is available to translate it into the relatively ugly Fortran version on the left.)

The point of these remarks is that the careful structuring of a program is very much facilitated if our programming language has just the right constructs -- not too few, but also not too many. Once again, understanding and correctness proofs are made much easier.

Incidentally, there are other ways in which languages can be improved, especially for the purposes of numerical computation. The only one we have worked on to any extent is the provision of variable precision arithmetic. A preprocessor has been developed to implement this facility [10]. The need for such a facility with ordinary differential equations arises during the CALCULATION stage with variable order methods, because, if such methods take their first couple of steps with very low order methods, they may need to use higher precision calculations during those steps when the tolerance is small.

I believe that persons interested in numerical computation should be more insistent with language designers about exactly what is required for numerical computation. Otherwise we will continue to get what some numerically illiterate compiler writer thinks we ought to have, which, under the circumstances, may be exactly what we deserve.

Comparisons

Comparing methods has been one of our main interests at the University of Toronto. In a paper with Enright, Fellen and Sedgwick [7], an attempt was made to put such comparisons on a reasonably firm basis. Problems, methods and criteria were carefully defined, and extensive experiments were performed. Conclusions were drawn regarding the relative efficiency and reliability of Runge-Kutta, variable order Adams, and extrapolation methods.

The testing program, called DETEST, has been modified slightly and rewritten in a well-structured form. A description of this program, and how to use it, is now almost ready to be published as a technical report [4].

During the earlier experiments we realized that some of the methods could be improved. We were also aware of the fact that our tinkering with other people's methods in order to make them conform to the requirements of DETEST might have degraded their performance to some extent. We are therefore writing our own well-structured versions of what appear to be the best methods. As already mentioned, some have recently been completed and are being prepared for publication.

In making decisions about how to structure programs, we have always maintained we would be willing to sacrifice at least small amounts of efficiency for the sake of such advantages. But it has turned out so far that the only decreases in efficiency have been quite negligible.

Our published results have been concerned only with non-stiff systems, and little attention has been given to difficulties such as coping with discontinuities. We have however been carrying out experiments with methods for stiff systems and a report is being prepared. We hope to do more about discontinuities at a later date.

Correctness

No amount of testing will prove that a program is correct in any worthwhile sense. Testing will help us find bugs, and testing will help us compare methods for reliability and efficiency, but something more is needed to prove correctness.

We must first decide what it is we want to prove. In fact, there are many senses in which we might like to prove a program correct, and we may sometimes prefer to speak about proving properties of a program.

One of the most important senses in which we might want to prove a program correct is to show that, for a particular class of problems, the program will always produce an approximation to the solution at XEND with the following property:

> there exists a function $z(x)$ which interpolates the initial value and the computed approximation, and which satisfies
>
> $$\|z'-f(x,z)\| < c\tau$$
>
> where c is some known constant, and τ is the tolerance.

We have called such theorems "effectiveness theorems". The first example was presented in [5], others in [2, 6, 11].

I hope that further effectiveness theorems will be developed, but that they will be tied more specifically to particular programs. Other properties need to be proven as well, one of the most important being a proof that a particular program will "fail-safe", in some worthwhile sense.

Certification, distribution, etc.

Even if we have developed well-structured programs that are known to be reliable and efficient, and, in some sense or other, correct, we are far from finished. The packaging of such programs, and making them available to many users on a variety of machines, is still an enormous task. It involves further testing, documentation, coordination of efforts, etc.

Fortunately, interest in this sort of activity is growing, as is evidenced by, for example, the NATS project [1] in North America and the NAG project [3] in Britain. It is to be hoped that these projects will receive very strong support. (Is it too much to hope for more coordination of the efforts between these two groups?)

Concluding remarks

I would like to conclude with a few remarks about current trends in the development of mathematical software. It seems to me that those of us interested in numerical computation are on the threshold of a major accomplishment, and we ought to put a considerable effort into seeing that it is carefully done.

The well-known contributions by Wilkinson and others in the field of linear algebra are at last being "packaged" [1] and made generally available. Other packages are being developed, and we should try to ensure that the high standards already established in linear algebra are maintained. We should insist on programs that are easy to understand and use, and that are known to be reliable, efficient and correct.

To contribute to this development, more numerical mathematicians than is presently the case need to take an interest in such areas as programming languages, and the packaging of programs for general use. There is a need to develop better interfacing between the user and the subroutines we develop. As the subroutines become better understood and increasingly dependable, we can give more attention to how they are used, and for what purposes they are used, and this in turn might have a considerable effect on how we teach about numerical methods.

Bibliography

1. James M. Boyle, William J. Cody, Wayne R. Cowell, Burton S. Garbow, Yasuhiko Ikebe, Cleve B. Moler, Brian T. Smith, "NATS, A Collaborative Effort to Certify and Disseminate Mathematical Software", Proceedings, ACM Conference (1972), pp. 630-635.
2. W.H. Enright, "Studies in the Numerical Solution of Stiff Ordinary Differential Equations", Ph.D. thesis, Department of Computer Science, University of Toronto (1972).
3. B. Ford, "The Nottingham Algorithms Group (NAG) Project", SIGNUM Newsletter, 8, 2 (April, 1973), pp. 16-21.
4. G. Hall, W.H. Enright, T.E. Hull, A.E. Sedgwick, "DETEST: A Program for Comparing Numerical Methods for Ordinary Differential Equations", Technical Report, Department of Computer Science, University of Toronto, in preparation.

5. T.E. Hull, "The Numerical Integration of Ordinary Differential Equations", IFIP Congress 68, Proceedings, (North-Holland, Amsterdam, 1968), pp. 131-144.

6. T.E. Hull, "The Effectiveness of Numerical Methods for Ordinary Differential Equations", SIAM Studies in Num Anal 2 (1970), pp. 114-121.

7. T.E. Hull, W.H. Enright, B.M. Fellen, A.E. Sedgwick, "Comparing Numerical Methods for Ordinary Differential Equations", SIAM J Num Anal, 9, 4 (December, 1972), pp. 603-637.

8. T.E. Hull, W.H. Enright, A.E. Sedgwick, "The Correctness of Numerical Algorithms", Proceedings of the SIGPLAN Symposium on Proofs of Assertions about Programs, Las Cruces, New Mexico (1972), pp. 66-73.

9. T.E. Hull, W.H. Enright, A.E. Sedgwick, "The Structure of Programs for Ordinary Differential Equations", Technical Report, Department of Computer Science, University of Toronto, in preparation.

10. T.E. Hull and J.J. Hofbauer, "Language Facilities for Numerical Computation", Technical Report, Department of Computer Science, University of Toronto, in preparation.

11. Arthur Sedgwick, "An Effective Variable Order Variable Step Adams Method", Ph.D. thesis, Department of Computer Science, University of Toronto (1973).

12. A.E. Sedgwick and C.A. Steele, "DEFT: A Disciplined Extension of FORTRAN", Technical Report, Department of Computer Science, University of Toronto, in preparation.

BOUNDARY CONDITIONS FOR HYPERBOLIC DIFFERENTIAL EQUATIONS

Heinz-Otto Kreiss

1. Systems in one space dimension.

In this chapter we collect some well known results for problems in one space dimension. Consider a hyperbolic system

(1.1) $$\partial u/\partial t = A\partial u/\partial x,$$

Here $u(x,t) = (u^{(1)}(x,t),\dots,u^{(n)}(x,t))^T$ denotes a vector function and A a constant $n \times n$ matrix. Hyperbolicity implies that A can be transformed to real diagonal form, i.e. there is a nonsingular transformation S such that

(1.2) $$SAS^{-1} = \begin{pmatrix} A^{I} & 0 \\ 0 & A^{II} \end{pmatrix} = \tilde{A}$$

where

$$-A^{I} = -\begin{pmatrix} a_1 & 0 & \dots & 0 \\ 0 & a_2 & & 0 \\ \vdots & & \ddots & \vdots \\ 0 & \dots & 0 & a_r \end{pmatrix} > 0, \quad A^{II} = \begin{pmatrix} a_{r+1} & 0 & \dots & 0 \\ 0 & a_{r+2} & & 0 \\ \vdots & & \ddots & \vdots \\ 0 & \dots & 0 & a_n \end{pmatrix} > 0$$

are positive definite diagonal matrices. We can thus introduce new variables

(1.3) $$v = Su$$

and get

$$(1.4) \qquad \partial v/\partial t = \tilde{A}\, \partial v/\partial x .$$

The last equation can also be written in partitioned form

$$(1.5) \qquad \partial v^{I}/\partial t = A^{I} \partial v^{I}/\partial x, \quad \partial v^{II}/\partial t = A^{II} \partial v^{II}/\partial x,$$

where $v^{I} = (v^{(1)},\ldots,v^{(r)})^{T}$, $v^{II} = (v^{(r+1)},\ldots,v^{(n)})^{T}$. (1.5) represents n scalar equations. Therefore we can write down its general solution:

$$(1.6) \qquad v^{(j)}(x,t) = v^{(j)}(x+a_j t), \quad j = 1,2,\ldots n,$$

which are constant along the characteristic lines $x + a_j t = \text{const.}$.

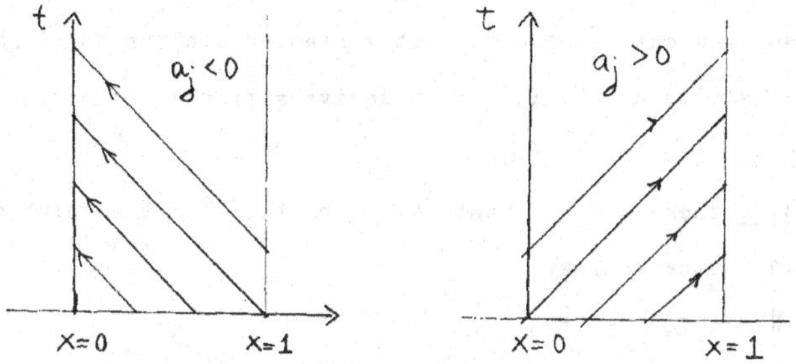

fig.1

Now consider (1.4) in the strip $1 \geq x \geq 0, \infty > t \geq 0$. The solution is uniquely determined and can be computed explicitly if we specify initial conditions

$$(1.8) \qquad v(x,0) = f(x), \quad 0 \leq x \leq 1,$$

and boundary conditions

$$(1.9) \qquad v^{II}(0,t) = R_0 v^{I}(0,t) + g_0(t), \quad v^{I}(1,t) = R_1 v^{II}(1,t) + g_1(t).$$

Here R_0, R_1 are rectangular matrices and g_0, g_1 are given vector functions. If we consider wave propagation then the boundary conditions describe how the waves are reflected at the boudary.

Nothing essentially is changed if $A = A(x,t)$ and $R_j = R_j(t)$ are functions of x,t. Now the characteristics are not straight lines but the solutions of the ordinary differential equations

$$dx/dt = a_j(x,t).$$

More general systems

$$(1.10) \quad \partial v/\partial t = \tilde{A}(x,t)\partial v/\partial x + B(x,t)v + F(x,t),$$

can be solved by the iteration

$$(1.11) \quad \partial v^{[n+1]}/\partial t = \tilde{A}(x,t)\partial v^{[n+1]}/\partial x + F^{[n]}$$

where

$$F^{[n]} = B(x,t)v^{[n]} + F.$$

Furthermore, it is no restriction to assume that A has diagonal form. If not, we can by a change of dependent variables achieve the form (1.10). There is also no difficulty to derive a priori estimates. One can show

<u>Theorem 1.1.</u> There are constants K, α such that for the solutions of (1.8)-(1.10) the estimate

$$(1.12) \quad \begin{aligned} \|u(x,t)\| &\le K e^{\alpha t}\|f(x)\| + \\ & K\,\ell(\alpha,t)\Big\{\max_{0\le\xi\le t}\|F(x,\xi)\| + \sum_{i=0}^{1}\max_{0\le\xi\le t}|g_i(\xi)|\Big\} \end{aligned}$$

holds. Here

$$\|u\| = \Big(\int_0^1 |u|^2 dx\Big)^{1/2}$$

denotes the usual L_2 - norm and $l(\alpha,t)$ is the function

$$\ell(\alpha,t) = \begin{cases} (1-e^{\alpha t})/\alpha & \text{if } \alpha \neq 0, \\ t & \text{if } \alpha = 0. \end{cases}$$

We can therefore develop a rather complete theory for initial boundary value problems by using characteristics. This has of course been known for a long time. The only trouble is, that this theory cannot be easily generalized to problems in more than one space dimension. For difference approximations it is already in one space dimension not adequate.

2. A simple example.

In this chapter we consider the system

$$(2.1)\quad \frac{\partial u}{\partial t} = \begin{pmatrix} -1 & 0 \\ 0 & +1 \end{pmatrix} \frac{\partial u}{\partial x} + \begin{pmatrix} 0 & 1 \\ 1 & 0 \end{pmatrix} \frac{\partial u}{\partial y}, \quad u = \begin{pmatrix} u^{(1)} \\ u^{(2)} \end{pmatrix}$$

We consider first the half planeproblem, i.e. we consider (2.1) for $\infty > x \geq 0$, $\infty > y > -\infty$, $\infty > t \geq 0$. For $t = 0$ we describe initial conditions

$$(2.2)\quad u(x, y, 0) = f(x)$$

and for $x = 0$ boundary conditions

$$(2.3)\quad u^{(1)}(0, y, t) = a\, u^{(2)}(0, y, t) + g(y, t).$$

Here a is a given complex number. We want to investigate for which values of a the above problem is well posed.

Connected with the initial boundary problem is a set of eigenvalue problems on the interval $\infty > x \geq 0$ which depend on a real parameter ω.

$$(2.5)\quad s\varphi = \begin{pmatrix} -1 & 0 \\ 0 & +1 \end{pmatrix} \frac{d\varphi}{dx} + i\omega \begin{pmatrix} 0 & 1 \\ 1 & 0 \end{pmatrix} \varphi, \quad \varphi = \begin{pmatrix} \varphi^{(1)} \\ \varphi^{(2)} \end{pmatrix},$$

with boundary conditions

$$(2.6)\quad \varphi^{(1)}(0) = a\, \varphi^{(2)}(0), \qquad \|\varphi\|^2 = \int_0^\infty |\varphi|^2 dx < \infty$$

S. Agmon [1] proved the following lemma.

Lemma 2.1. The initial boundary value problem is not well posed if the eigenvalue problem has for some $\omega = \omega_0$ a nontrivial solution $s = s_0$ with Real $s > 0$.

Proof: Let $\varphi(x)$ be an eigenfunction corresponding to the eigenvalue $s = s_0$. Then

$$u_\alpha(x,y,t) = e^{\alpha(s_0 t + i\omega_0 y)} \varphi(\alpha x)$$

is a solution of (2.1) with initial values $f(x) = \varphi(\alpha x)$ and homogeneous boundary conditions $u_\alpha^{(1)}(0,y,t) = a\, u_\alpha^{(2)}(0,y,t)$. Here $\alpha > 0$ can be any positive constant. Therefore (2.1)-(2.3) has solutions which grow arbitrarily fast with time.

A simple calculation, already performed by R.Hersch [2], gives

<u>lemma 2.2.</u> The eigenvalue problem (2.5)-(2.6) has an eigenvalue s with Real s > 0 if and only if a is not real and $|a| > 1$.

<u>Proof:</u> Let $\omega \neq 0$. The general solution of (2.5) is given by

$$\varphi = \sigma_1 \begin{pmatrix} \frac{i\omega}{s+\varkappa} \\ 1 \end{pmatrix} e^{\varkappa x} + \sigma_2 \begin{pmatrix} \frac{i\omega}{s-\varkappa} \\ 1 \end{pmatrix} e^{-\varkappa x}, \quad \varkappa = \sqrt{s^2+\omega^2}$$

For (2.6) to be fulfilled the following conditions are necessary and sufficient.

$$\sigma_1 = 0 \quad , \quad i\omega = (s-\varkappa)a .$$

These conditions cannot be fulfilled if $|a| < 1$. If $|a| \geq 1$ then there is a solution with

$$(2.7) \qquad s = \frac{a^2+1}{2ai}\,\omega, \qquad \varkappa = \sqrt{\left(\frac{a^2+1}{2ai}\right)^2 (1+\omega^2)} .$$

If $|a| = 1$ or if a is real then $(a^2 + 1)/2ai$ is purely imaginary and s is also purely imaginary. If $|a| > 1$ and a is not real then there is an eigenvalue s with Real s > 0. This proves the lemma.

For later purposes it is essential to investigate also the eigenvalues with Real s = 0. If $|a| < 1$ then there are no eigenvalues for Real $s \geq 0$. In fact there is a constant $\delta > 0$ such that

$$(2.8) \qquad |i\omega - (s-\varkappa)a| \geq \delta(|s| + |\omega|)$$

If $|a| = 1$, a not real, then by (2.7) there is an eigenvalue and its corresponding eigenfunction is given by

$$(2.9)\qquad \varphi = \begin{pmatrix} \frac{i\omega}{s-\varkappa} \\ 1 \end{pmatrix} e^{-\varkappa x}, \quad s = \frac{a^2+1}{2ai}\,\omega, \quad \varkappa = \sqrt{s^2+\omega^2}$$

Here Real $\varkappa \geq$ const.$(|s| + |\omega|)$.If $a = \pm 1$ then (2.9) is also a solution.However,$\varkappa = 0$ and therefore φ does not belong to L_2. We call $s = \mp i\omega$ a generalized eigenvalue.Finally,if a is real and $|a| > 1$ then Real $\varkappa = 0$ and (2.9) represents again a generalized eigenvalue.We have therefore proved

<u>Lemma 2.3</u>.For $|a| < 1$ there are no eigenvalues or generalized eigenvalues with Real $s \geq 0$.For $|a| = 1$,a not real,there are eigenvalues with Real $s = 0$ but no generalized eigenvalues.For $|a| \geq 1$, a real,there are generalized eigenvalues with Real $s = 0$ but no eigenvalues.

We shall now investigate the behavior of (2.1)-(2.3) for $|a| \leq 1$ and for $\infty > a > -\infty$ in detail.For simplicity we assume that the initial values are homogeneous,i.e.$f(x) = 0$.This is no restriction. If $f(x) \not\equiv 0$ then we solve the appropriate Cauchy problem first.We construct the solution explicitly. Fourier transform (2.1)-(2.3) with respect to y.Let ω denote the (real) dual variable and denote by $\hat{u} = \mathcal{F} u$ the Fourier transform of u.Then

$$(2.10)\qquad \begin{aligned} \frac{\partial \hat{u}}{\partial t} &= \begin{pmatrix} -1 & 0 \\ 0 & +1 \end{pmatrix} \frac{\partial \hat{u}}{\partial x} + i\omega \begin{pmatrix} 0 & 1 \\ 1 & 0 \end{pmatrix} \hat{u}, \\ \hat{u}(x,\omega,0) &= 0, \\ \hat{u}^{(1)}(0,\omega,t) &= a\,\hat{u}^{(2)}(0,\omega,t) + \hat{g}(\omega,t). \end{aligned}$$

(2.10) represents a set of one dimensional problems which by the preceding chapter have unique solutions.Therefore they can be solved by Laplace transform in time.Let

$$v = v(x,\omega,s) = \mathcal{L}\hat{u} = \int_0^\infty e^{-st}\hat{u}\,dt, \quad h = \mathcal{L}\hat{g}, \quad \text{Real } s > 0,$$

then

$$(2.11)\quad s v = \begin{pmatrix} -1 & 0 \\ 0 & +1 \end{pmatrix} \frac{dv}{dx} + i\omega \begin{pmatrix} 0 & 1 \\ 1 & 0 \end{pmatrix} v ,$$

$$(2.12)\quad v^{(1)}(0,\omega,s) = a\, v^{(2)}(0,\omega,s) + h, \quad \int_0^\infty |v|^2 dx < \infty .$$

(2.11) is a system of ordinary differential equations.Its general solution has the form

$$v = \sigma_1 \begin{pmatrix} \frac{i\omega}{s+\varkappa} \\ 1 \end{pmatrix} e^{\varkappa x} + \sigma_2 \begin{pmatrix} \frac{i\omega}{s-\varkappa} \\ 1 \end{pmatrix} e^{-\varkappa x} ,$$

where $\varkappa = \pm\sqrt{s^2+\omega^2}$ are the solutionsof the characteristic equation

$$(2.13)\quad \varkappa^2 - (s^2+\omega^2) = 0$$

The boundary conditions are fulfilled if

$$\sigma_1 = 0 , \qquad \sigma_2 = \frac{s-\varkappa}{i\omega - (s-\varkappa)a}\, h .$$

Let $s = i\xi + \eta$, ξ, η real.Inverting the Fourier transform and Laplace transform gives us

$$(2.14)\quad u(x,y,t) = (2\pi)^{-1} e^{\eta t} \int_{-\infty}^{+\infty}\int_{-\infty}^{+\infty} e^{-i\xi t - i\omega y} v(x,\omega,i\xi+\eta)\, d\xi\, d\omega .$$

Here $\eta > 0$ is an arbitrary (chosen)fixed positive constant. Parseval's relation gives us the following estimate

$$N(u,\eta) = \int_{-\infty}^{+\infty}\int_0^\infty\int_0^\infty e^{-2\eta t} |u(x,y,t)|^2 dt\, dx\, dy =$$

$$(2.15)\quad \int_{-\infty}^{+\infty}\int_0^\infty\int_{-\infty}^{+\infty} |v(x,\omega,i\xi+\eta)|^2 d\xi\, dx\, d\omega =$$

$$\int_{-\infty}^{+\infty}\int_0^\infty\int_{-\infty}^{+\infty} |\sigma_2|^2 \left(\left|\frac{i\omega}{s-\varkappa}\right|^2 + 1 \right) e^{-(\text{Real }\varkappa)x} d\xi\, dx\, d\omega =$$

$$\int_{-\infty}^{+\infty}\int_{-\infty}^{+\infty} \frac{g(\omega', s')}{\text{Real }\varkappa} |h|^2 d\omega\, d\xi ,$$

where
$$\rho(\omega',s') = \frac{|\omega'|^2 + |s' - \varkappa'|^2}{|i\omega' - (s'-\varkappa')a|^2}$$

$$\omega' = \omega/\tau\,,\quad s' = s/\tau\,,\quad \varkappa' = \varkappa/\tau\,,\quad \tau = |\omega| + |s|\,.$$

Consider now the case that $|a| < 1$. Observing that $|\text{Real}\ \varkappa'| \geq \eta$ it follows from (2.8) that $\rho(\omega',s') \leq 2\delta^{-2}$ and therefore

$$N(u,\eta) \leq 2\delta^{-2}\eta^{-1}\int_{-\infty}^{\infty}\int_{0}^{\infty}|h|^2\,d\omega\,d\xi = \tag{2.16}$$
$$2\delta^{-2}\eta^{-1}\int_{-\infty}^{\infty}\int_{0}^{\infty}e^{-2\eta t}|g(y,t)|^2\,dt\,dy.$$

We can therefore estimate the solution in terms of the boundary data.

Let $|a| = 1$, a not real. Then $\rho(\omega',s')$ is bounded except in a neighbourhood of the eigenvalue $s' = (a^2+1)\omega'/2ai$. Here $\rho(\omega',s') \sim$ const.$(|\omega| + |s|)/\eta$. Thus the best estimate we can achieve is

$$N(u,\eta) \leq \text{const.}\,\eta^{-1}\int_{-\infty}^{\infty}\int_{-\infty}^{\infty}(|\omega|+|s|)|h|^2\,d\omega\,d\xi \leq \tag{2.17}$$
$$\text{const}\,\eta^{-1}\int_{-\infty}^{\infty}\int_{0}^{\infty}e^{-2\eta t}(|g|^2 + |\partial g/\partial t|^2 + |\partial g/\partial y|^2)\,dt\,dy.$$

Therefore we cannot estimate u with g alone but we have to add the first derivatives of g as well. However, if we are only interested in interior estimates, i.e. we want to estimate u for $x \geq \varepsilon > 0$, then we again can estimate u with help of g alone. The reason is that v decays like $\exp(-|\omega|x)$ in the x direction.

Similar estimates hold for the case that $a = \pm 1$. Though $s = i\omega$ is a generalized eigenvalue of (2.5)-(2.6) we are saved by the fact that $\varkappa = 0$ is a double root of the characteristic equation (2.13). Therefore if $s = i\omega + \eta$ then $\varkappa = \sqrt{i\omega\eta}$ and v decays like $\exp(-(|\omega|x)^{\frac{1}{2}})$. We call $s = i\omega$ a generalized eigenvalue of the first kind.

Now let $|a| > 1$, a real. $\rho(\omega',s')$ is bounded except in a neighbourhood of the generalized eigenvalue $s = (a^2+1)\omega/2ai$. Here

$g(|\omega'|,|s'|) \sim (|\omega|+|s|)^2/\eta^2$. Furthermore Real $\varkappa \sim \eta$. Therefore the best estimate we can get is

$$N(u,\eta) \le \text{const}\,\eta^{-3}\int_{-\infty}^{\infty}\int_{-\infty}^{\infty}(|\omega|+|s|)^2\cdot|\hat{h}|^2 d\omega d\xi \tag{2.18}$$
$$\le \text{const}.\,\eta^{-3}\int_{-\infty}^{\infty}\int_{0}^{\infty}e^{-2\eta t}(|g|^2+|\partial g/\partial t|^2+|\partial g/\partial y|)dt\,dy.$$

What is worse we cannot get any better interior estimate because $v \sim \exp(-\eta x)$ does not decay fast enough. In this case we call $s = (a^2+1)\omega/2ai$ a generalized eigenvalue of the second kind. The situation is very delicate especially if we have more than one boundary involved. Consider for example (2.1) in the strip $1 \ge x \ge 0, t \ge 0$, $\infty > y > -\infty$. For $x = 1$ we describe similar boundary conditions as for $x = 0$. In this case there are solutions which lose more and more derivatives with time. These are waves which are reflected back and forth between the boundaries losing derivatives every time they are reflected. In this case the corresponding eigenvalue problem has solutions $s = \text{const.}\log|\omega|$.

We summarize our results in

<u>Theorem 2.1.</u> An estimate of type (2.16) holds if and only if the eigenvalue problem (2.5),(2.6) has no eigenvalue or generalized eigenvalue for Real $s \ge 0$. Interior estimates of the same type hold if there are no eigenvalues for Real $s > 0$ and for Real $s = 0$ there are only eigenvalues or generalized eigenvalues of the first kind. If there are generalized eigenvalues of the second type then one cannot estimate u without using derivatives of g and one can lose more and more derivatives with time if other boundaries are present.

3. Problems in more than one space dimension.

One can generalize the results of the last chapter to rather general hyperbolic systems

$$(3.1)\qquad \partial u/\partial t = P(\partial/\partial x)\,u$$

where

$$P(\partial/\partial x) = A\,\partial/\partial x_1 + \sum_{j=2}^{m} B_j\,\partial/\partial x_j$$

Here A, B_j are constant sqare matrices of order n.Without restriction we can assume that A has the diagonal form (1.2).We consider first the half-planeproblem,i.e. we consider (3.1) for $x_1 \geq 0, \infty > x_j > -\infty$, $j = 2,\ldots,m$, $t \geq 0$ and give homogeneous initial values

$$(3.2)\qquad u(x,0) = 0$$

for t = 0 and boundary conditions

$$(3.3)\qquad u^{I}(0,x_-,t) = R\,u^{II}(0,x_-,t) + g(x_-,t),\quad x_- = (x_2,\ldots,x_n)^T$$

for x = 0.The associated eigenvalue problem is now given by

$$(3.4)\qquad s\varphi = A\,d\varphi/dx + i\,B(\omega)\varphi,\quad B(\omega)\varphi = \sum_{j=2}^{m} \omega_j B_j$$

$$(3.5)\qquad \varphi^{I}(0) = R\,\varphi^{II}(0),\quad \|\varphi\| < \infty .$$

One can show that lemma 2.1 and theorem 2.1 also hold in this general case.Details will be given in a forthcoming paper where also a number of examples , the Shallow Water Equations and Maxwell's equations, will be treated.

The theory for equations with variable coefficients in domains with curved but smooth boundaries has been developed completely only for the case that there are no eigenvalues or generalized eigenvalues for Real $s \geq 0$ [4],[5],[6]. The case of generalized eigenvalues will be treated in [3].

The theory for difference approximation is completely analogous and will be discussed in a forthcoming paper. However, it is technically and notationally much more complicated.

References

1. Agmon,S.,Report,Paris Conference on Partial Differential Equations, 1962.

2. Hersh,R.,Mixed problems in several variables,J.Math.Mech.,Vol.12, 1963.

3. Elvius,T.,Kreiss,H.-O.Initial boundary value problems for hyperbolic systems II,to appear.

4. Kreiss,H.-O,Initial boundary value problems for hyperbolic systems, Comm.Pure Appl.Math.,Vol.23,1970.

5. Sakamoto,R.,J.Math.Kyoto Univ.Vol 10,1970.

6. Agranovic,M.S.,Boundary value problems for systems with a parameter, Mat.Sbornik USSR,Vol.13,1971.

NONLINEAR METHODS FOR STIFF SYSTEMS OF ORDINARY DIFFERENTIAL EQUATIONS

J. D. Lambert

I. INTRODUCTION

For the purposes of this paper, a linear method for the numerical solution of the initial value problem (IVP) $y' = f(x,y)$, $y(a) = \eta$, is defined to be a method which, when applied to the test equation $y' = Ay$, A a dense matrix, yields a linear difference equation in the discrete variable y_n. Thus the well-known classes of linear multistep (LM), predictor-corrector (PC), and Runge-Kutta methods (RK) are all linear in this sense. Such methods are supported by substantial analysis and computational experience, and, normally, there is no reason to consider more bizarre classes of method which are nonlinear in the sense of the above definition. (For specialized applications of nonlinear methods, see [4], [5], [6]). However, the performance of LM, PC, and RK methods when applied to stiff systems can hardly be said to be satisfactory. Such linear methods can possess adequate stability properties to cope with stiff systems (i.e. are A- or A(α)- stable) only if they are implicit. Moreover, when the system is stiff, the resulting implicit system of difference equations cannot be solved satisfactorily by direct iteration, and some form of Newton iteration, with the resultant need to calculate inverses of Jacobians, is necessary. This is the real computational problem with stiff systems. It is the purpose of this paper to investigate the possibility of constructing explicit nonlinear methods which have adequate stability to enable them to handle stiff systems.

The mechanism we shall use to construct such methods is that of [4], namely local representation of the solution by a rational function. (The particular methods constructed in [4] were developed to deal with singularities, and are quite unsuitable for stiff systems.) By way of motivation, consider the problem of polynomial interpolation at $x = \frac{1}{2}$ of the data given by the function 4^x at $x = 0,1,2,3,\ldots$. One readily finds that the most acceptable answer is given by linear interpolation between the points x = 0 and x = 1. Attempts to interpolate with higher degree polynomials, or to extrapolate, yield disastrous results. Since the components of the solution of a stiff system contain rapidly decaying exponentials, and since the application of a LM method is equivalent to local representation of the solution by a polynomial, the above remarks on the interpolation problem mirror, in a simplified way, the well-known results of Dahlquist [1] and Widlund [8] which limit the order of implicit LM methods, and rule out explicit LM methods, if the condition of A- or A(α)-stability is to be met. The difficulty in the interpolation problem is overcome if we switch from polynomial to rational interpolation. We are therefore motivated to construct methods which are related to local rational interpolation in the sense that LM methods are related to local polynomial interpolation.

II. AN ELEMENTARY NONLINEAR METHOD

Let the theoretical solution, $y(x)$, of the scalar IVP $y' = f(x,y)$, $y(a) = \eta$, be represented locally in $[x_n, x_{n+1}]$ by the rational function $I(x) = A/(x+B)$. If y_n is an approximation to $y(x_n)$ and $f_n = f(x_n, y_n)$, where $x_n = a + nh$, we impose the requirements

$$y_n = I(x_n), \quad y_{n+1} = I(x_{n+1}), \quad f_n = I'(x_n).$$

On eliminating A and B from these three equations we obtain the method

$$y_{n+1} - y_n = \frac{hy_n f_n}{y_n - hf_n} \tag{1(i)}$$

(A similar process applied with $I(x) = Ax + B$ yields Euler's Rule.) We immediately have to impose the restriction that

$$|y(x)| + |y'(x)| \neq 0, \quad x \geq a. \tag{1(ii)}$$

If it happens that, despite (1(ii)), $y_n - hf_n$ vanishes for a particular h, then another value for h must be chosen.

<u>Applicability to systems.</u>

Method (1) is <u>component-applicable</u> to the system $\underset{\sim}{y}' = \underset{\sim}{f}(x, \underset{\sim}{y})$ in the sense that if $\underset{\sim}{y} = [{}^1y, {}^2y, \ldots, {}^my]^T$, $\underset{\sim}{f} = [{}^1f, {}^2f, \ldots, {}^mf]^T$, then we may compute with the method

$$ {}^{i}y_{n+1} - {}^{i}y_{n} = \frac{h\,{}^{i}y_{n}\,{}^{i}f_{n}}{{}^{i}y_{n} - h\,{}^{i}f_{n}} \qquad i = 1,2,\ldots,m \qquad (1(iii)) $$

Stability.

Applying (1) to the test equation $\underset{\sim}{y}' = \lambda\underset{\sim}{y}$, $\mathrm{Re}\lambda < 0$, we find

$$ {}^{i}y_{n+1}/{}^{i}y_{n} = 1/(1-h\lambda) \qquad i = 1,2,\ldots,m. $$

Thus ${}^{i}y_{n+1}/{}^{i}y_{n}$ is the (0,1) Padé approximation to $e^{h\lambda}$. The method is consequently A-stable : indeed, it is L-stable (see [2]). (Note that (1), applied to the system $\underset{\sim}{y}' = \lambda\underset{\sim}{y}$, thus yields a linear difference system; applied to $\underset{\sim}{y}' = A\underset{\sim}{y}$, A a dense matrix, however, it yields a nonlinear difference system, and is thus nonlinear in the sense of section I.)

It is important to observe that the test equation $\underset{\sim}{y}' = \lambda\underset{\sim}{y}$ is, as far as linear methods are concerned, essentially equivalent to the test equation $\underset{\sim}{y}' = A\underset{\sim}{y}$, where A is a dense matrix whose eigenvalues are distinct and lie in the left half-plane. This is not the case for the methods of this paper. For example, if we make the transformation $\underset{\sim}{y} = H\underset{\sim}{z}$, where $H^{-1}AH = \Lambda = \mathrm{diag}(\lambda_1,\lambda_2,\ldots,\lambda_m)$, then the system $\underset{\sim}{y}' = A\underset{\sim}{y}$ is transformed into $\underset{\sim}{z}' = \Lambda\underset{\sim}{z}$. Euler's rule, applied to the original system yields $\underset{\sim}{y}_{n+1} = (I + hA)\underset{\sim}{y}_n$, which, on applying the transformation $\underset{\sim}{y}_n = H\underset{\sim}{z}_n$, becomes $\underset{\sim}{z}_{n+1} = (I + h\Lambda)\underset{\sim}{z}_n$, which is Euler's rule applied to the transformed system. Similar statements do not hold for the method (1), and for such methods our analysis is necessarily restricted to the test equation $\underset{\sim}{y}' = \Lambda\underset{\sim}{y}$, which is essentially equivalent to the scalar test equation $y' = \lambda y$. (In the remainder of this paper we shall consider only the scalar test equation.) Partial, and inconclusive, results do exist for the full system. Thus,

Theorem 1 Let method (1(iii)) be applied to $\underset{\sim}{y}' = A\underset{\sim}{y}$, A a real matrix with distinct eigenvalues $\lambda_1,\lambda_2,\ldots,\lambda_m$, and corresponding eigenvectors $\underset{\sim}{c}_1,\underset{\sim}{c}_2,\ldots\underset{\sim}{c}_m$. Then the resulting (nonlinear) difference system has m independent solutions

$$ \underset{\sim}{y}_{n,i} = \left(\frac{1}{1-h\lambda_i}\right)^{n}\underset{\sim}{c}_i\,, \qquad i = 1,2,\ldots,m. $$

The inconclusiveness of this result arises from the fact that knowledge of m independent solutions of a nonlinear m-dimensional system of difference equations does not enable us to construct the general solution.

Order and local truncation error.

We associate with the method (1) the nonlinear operator $P[y(x);h]$, defined by

$$P[y(x);h] = y(x+h)-y(x) - hy(x)y'(x)/[y(x)-hy'(x)]$$

where $y(x)$ is an arbitrary function in C^1 such that $|y(x)| + |y'(x)| \neq 0, \forall x$. If $P[y(x); h] = O(h^{p+1})$ we shall say that the method has order p. The local truncation error, T_{n+1}, at x_{n+1} is then defined to be $P[y(x_n); h]$, where $y(x)$ is now taken to be the theoretical solution of the IVP. It immediately follows, under the usual localizing assumption that $y_n = y(x_n)$, that $y(x_{n+1}) - y_{n+1} = T_{n+1}$.

It is obvious from (1(i)) that, independently of f, $y_n = 0$ implies $y_{n+1} = 0$. Thus the method fails to follow the solution through a zero. We can interpret this phenomenon by considering the local truncation error of (1). Expanding $P[y(x_n); h]$ about x_n, we find that

$$T_{n+1} = h^2[\tfrac{1}{2}y^{(2)} - y^{(1)^2}/y]_{x=x_n} + O(h^3) \qquad (1(iv))$$

indicating that the method has order 1 in general. It is tempting, but incorrect, to ascribe the failure of the method when $y_n = 0$ to the fact that the coefficient of h^2 in T_{n+1} is then infinite. (Recall that y_n' cannot also be zero; see (1(ii)).) Firstly, later examples occur in this paper of methods which behave perfectly well even when the coefficient of h^{p+1} appears to be infinite. Secondly, the above argument suggests that if $y_n = 0$, the local truncation error at x_n will be infinite, and this is clearly not the case; the numerical solution simply progresses along the x-axis, whilst the theoretical solution crosses it. However, if we expand numerator and denominator in $P[y(x_n); h]$ separately, we obtain, in place of (1(iv)),

$$T_{n+1} = \left.\frac{[yy^{(2)} - 2y^{(1)^2}]h^2/2! + [yy^{(3)} - 3y^{(1)}y^{(2)}]h^3/3! + O(h^4)}{y - hy^{(1)}}\right|_{x=x_n} \qquad (1(v))$$

It is now clear that $T_{n+1} = O(h^{p+1})$, $p \geq 1$ if $y_n \neq 0$, but that $T_{n+1} = O(h)$ if $y_n = 0$. Thus, if $y_n \neq 0$, the method has order at least 1, and if $y_n = 0$, it has order precisely zero, that is, is <u>locally inconsistent</u>. The actual behaviour of the numerical solution when $y_n = 0$ is just what we would expect from an inconsistent method. Our conclusion is that the notion of "principal" local truncation error, as typified by (1(iv)), is misleading for nonlinear methods, and we shall henceforth express local truncation errors in the form of (1(v)). We note also that the order of a nonlinear method is a function of the IVP, and can change as the solution progresses.

III. SOME HIGHER ORDER NONLINEAR METHODS

The fact that method (1) becomes inconsistent when $y_n = 0$ (despite the possibility of overcoming the difficulty in practice by applying local transformations of the form $y = \hat{y}$ + constant) motivates us to seek other methods not having this undesirable property. The local inconsistency of (1) can be interpreted in terms of the geometry of the underlying interpolant $I(x) = A/(x+B)$. This function can be zero only for infinite x, that is, on its horizontal asymptote; extrapolation using this part of the curve naturally produces a constant zero solution. We are thus naturally led to consider the local interpolant $I(x) = (Ax+B)/(x+C)$, which can be zero for finite x. Applying the conditions

$$y_{n+j} = I(x_{n+j}), \quad j = 0,1,2, \quad f_{n+1} = I'(x_{n+1})$$

yields the two-step nonlinear method

$$y_{n+2} - y_{n+1} = \frac{h(y_{n+1} - y_n)f_{n+1}}{2(y_{n+1} - y_n) - hf_{n+1}} \tag{2(i)}$$

whose local truncation error is

$$T_{n+2} = \left. \frac{[\frac{1}{3} y^{(1)}y^{(3)} - \frac{1}{2}y^{(2)^2}]h^3 + [\frac{1}{3} y^{(1)}y^{(4)} - \frac{2}{3} y^{(2)}y^{(3)}]h^4 + O(h^5)}{y^{(1)} - \frac{1}{6}h^2y^{(3)} - \frac{1}{12}h^3y^{(4)} + O(h^4)} \right|_{x=x_n} \tag{2(ii)}$$

It follows that the method has order at least 2 if $y_n^{(1)} \neq 0$, but has order 0 if $y_n^{(1)} = 0$, $y_n^{(2)} \neq 0$. The method thus fails to follow the solution through a maximum or minimum, a fact immediately deducible from (2(i)), since $y_{n+1} = y_n$ implies $y_{n+2} = y_{n+1}$, independently of f.

<u>Stability</u>.

Applying (2(i)) to the test equation $y' = \lambda y$, we obtain

$$y_{n+2} = y_{n+1} \frac{y_{n+1} - (1+\frac{1}{2}\bar{h})y_n}{(1-\frac{1}{2}\bar{h})y_{n+1} - y_n}, \qquad \bar{h} = h\lambda .$$

Putting $w_n = y_{n+1}/y_n$ gives

$$w_{n+1} = \frac{w_n - (1+\frac{1}{2}\bar{h})}{(1-\frac{1}{2}\bar{h})w_n - 1} \tag{2(iii)}$$

We now apply the following lemma, whose proof is trivial.

<u>Lemma 1</u>. Let w_n, $n = 0,1,2,\ldots$, a,b, and $c \in \mathbb{C}$, where $a^2 \neq -bc$.

Let

$$w_{n+1} = \frac{aw_n + b}{cw_n - a} \quad \text{if } w_n \text{ finite, and } w_{n+1} = \frac{a}{c} \text{ if } w_n \text{ infinite.}$$

Then

$$w_n = \begin{cases} w_0 & \text{if } n \text{ even} \\ w_1 & \text{if } n \text{ odd} \end{cases} .$$

Applying this lemma to (2(iii)), we have that

$$y_n/y_0 = \prod_{i=0}^{n-1} w_i = \begin{cases} (w_0 w_1)^{n/2} & \text{if } n \text{ even} \\ (w_0 w_1)^{(n-1)/2} w_0 & \text{if } n \text{ odd.} \end{cases}$$

Thus $y_n \to 0$ as $n \to \infty$ iff $|w_0 w_1| < 1$, that is, iff $|y_2/y_0| < 1$. In particular, if the additional starting value y_1 is obtained from the trapezoidal rule, $w_0 = y_1/y_0 = (1+\frac{1}{2}\bar{h})/(1-\frac{1}{2}\bar{h})$; on substituting in (2(iii)) we find that w_1 is also given by $(1+\frac{1}{2}\bar{h})/(1-\frac{1}{2}\bar{h})$. It then follows from the lemma that

$$\frac{y_{n+1}}{y_n} = \frac{1 + \frac{1}{2}\bar{h}}{1 - \frac{1}{2}\bar{h}} \qquad \text{for all } n ,$$

and the resulting method is then A-stable.

We return to the undesirable feature of method (2), namely its local inconsistency if $y_n^{(1)} = 0$, $y_n^{(2)} \neq 0$. We can interpret this in terms of the geometry of the interpolant $I(x) = (Ax+B)/(x+C)$ as follows. This function can have zero slope only for infinite x, that is, on its horizontal asymptote; it therefore (quite correctly) extrapolates to give $y_{n+2} = y_{n+1}$. What we would like the method to do, as the theor-

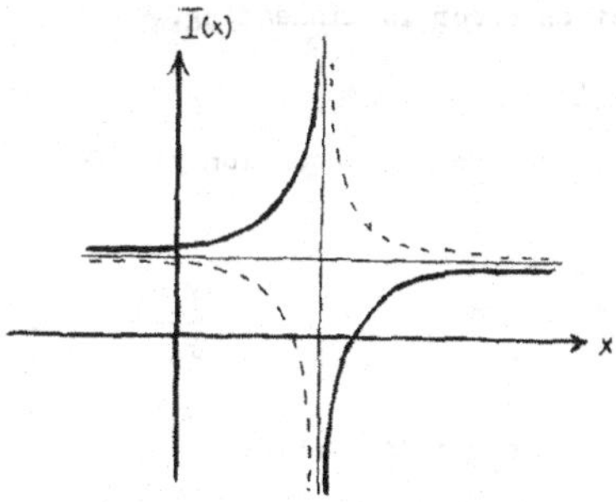

etical solution approaches and passes through, say, a maximum, is to extrapolate first with the function I(x) in the configuration indicated in the diagram by a solid line, and then, after the maximum has been passed, to extrapolate with I(x) in the configuration indicated by the broken line. Method (2) appears to carry insufficient information to enable this change of configuration to take place, and the extrapolation continues, after the maximum has been passed, to be made with I(x) in the configuration indicated by the solid line, whence the inconsistency. The information which is lacking is knowledge of the curvature of the solution. We are therefore motivated to construct a method based on the same local interpolant, namely $I(x) = (Ax+B)/(x+C)$, but now required to satisfy the conditions

$$y_{n+j} = I(x_{n+j}),\ j = 0,1,\quad f_n = I^{(1)}(x_n),\quad f_n^{(1)} = I^{(2)}(x_n)\ ,$$

where $f_n^{(1)} = f^{(1)}(x,y_n)$ and $f^{(1)}$ is the first total derivative of f with respect to x, obtained by differentiating the differential equation. The resulting one-step nonlinear method is

$$y_{n+1} - y_n = \frac{2hf_n^2}{2f_n - hf_n^{(1)}}\ ,\qquad |y^{(1)}(x)| + |y^{(2)}(x)| \neq 0 \qquad (3(i))$$

whose local truncation error is

$$T_{n+1} = \left.\frac{[-\frac{1}{2}y^{(2)^2} + \frac{1}{3}y^{(1)}y^{(3)}]h^3 + O(h^4)}{2y^{(1)} - hy^{(2)}}\right|_{x=x_n} \qquad (3(ii))$$

The method thus has order at least 2 if $y_n^{(1)} \neq 0$, and has order precisely 1 if $y_n^{(1)} = 0$; thus local inconsistency is avoided. Note that if we expand (3(ii)) in the manner of (1(iv)), we obtain

$$T_{n+1} = h^3\left[\frac{1}{6}y^{(3)} - \frac{1}{4}\frac{y^{(2)^2}}{y^{(1)}}\right]_{x=x_n} + O(h^4)$$

This representation suggests that the method will fail when $y_n = 0$. That this is not the case corroborates our earlier remarks that the concept of "principal" local truncation error is misleading.

Stability.

On applying the test equation $y' = \lambda y$, we easily find that

$$\frac{y_{n+1}}{y_n} = \frac{1+\frac{1}{2}\bar{h}}{1-\frac{1}{2}\bar{h}} \ ,$$

and the method is A-stable.

Having established the desirability of including information on the second derivative of the solution, we seek a higher order method using the local interpolant $I(x) = (Ax+B)/(x^2+Cx+D)$, and the conditions $y_{n+j} = I(x_{n+j})$, $j = 0,1,2$, $f_{n+1} = I^{(1)}(x_{n+1})$, $f^{(1)}_{n+1} = I^{(2)}(x_{n+1})$. The resulting two-step nonlinear method is

$$y_{n+2} = \frac{4y^2_{n+1}(y_{n+1}-y_n) - 4hy_{n+1}y_n f_{n+1} - h^2(2f^2_{n+1} - y_{n+1}f^{(1)}_{n+1})y_n}{4y_{n+1}(y_{n+1}-y_n) - 4hy_{n+1}f_{n+1} + h^2[2f^2_{n+1} - (y_{n+1}-2y_n)f^{(1)}_{n+1}]} \ , \qquad (4(i))$$

$$|y(x)| + |y^{(1)}(x)| + |y^{(2)}(x)| \neq 0 \ .$$

The local truncation error of this method is rather too complicated to quote here, but from it we may deduce that the method has order p, where

$$\begin{aligned} p &\geq 3, \text{ in general} \\ p &\geq 2, \text{ if } y^{(1)}_n = y^{(2)}_n = 0, \\ p &= 1, \text{ if } y_n = y^{(1)}_n = 0. \end{aligned} \qquad (4(ii))$$

Stability.

On applying method (4) to the test equation $y' = \lambda y$, and setting $w_n = y_{n+1}/y_n$ as before, we obtain, after some manipulation,

$$w_{n+1} = \frac{w_n - (1+\frac{1}{2}\bar{h})^2}{(1-\frac{1}{2}\bar{h})^2 w_n - (1-\frac{1}{2}\bar{h}^2)} \qquad (4(iii))$$

Lemma 1 cannot be applied to this equation. It is necessary to obtain a full solution of the nonlinear difference equation

$$w_{n+1} = \frac{aw_n + b}{cw_n + d} \ .$$

Clearly there exist two constant solutions of this equation, namely $w_n = \hat{w}$, where $\hat{w}$ is either root of the <u>characteristic polynomial</u> $cw^2 + dw = aw + b$. The following theorem and its corollaries enable us to investigate the behaviour of the solutions of (4(iii)).

<u>Theorem 2</u>. Let w_n, $n = 0,1,2,\ldots$, a, b, c, and $d \in \mathbb{C}$, where $ad \neq bc$, $c \neq 0$, and $b \neq 0$. Let

$$w_{n+1} = \frac{aw_n + b}{cw_n + d} \text{ if } w_n \text{ finite, and } w_{n+1} = \frac{a}{c} \text{ if } w_n \text{ infinite.}$$

Then
$$\left.\begin{aligned} w_n &= \hat{w}\,(1+\epsilon_n) \\ \epsilon_n &= \frac{A^n \epsilon_0}{C[\sum_{j=0}^{n-1} A^{n-j-1} D^j]\epsilon_0 + D^n} \end{aligned}\right\} \quad n = 0,1,\ldots,$$

where $\hat{w}$ is a root of $cw^2 + (d-a)w - b = 0$, and $A = a - c\hat{w}$, $C = c\hat{w}$, and $D = d + c\hat{w}$.

<u>Remark</u>. The above solution holds in all cases, no matter how often w_n becomes infinite.

<u>Corollary 1</u>. w_n is finite for $n = 0,1,\ldots,m-1$, and w_m is infinite if and only if the starting value $w_0 (= \hat{w}(1+\epsilon_0))$ is such that

$$\epsilon_0 = \frac{-D^m}{C \sum_{j=0}^{m-1} A^{m-j-1} D^j}$$

<u>Corollary 2</u>. Let the characteristic polynomial $cw^2 + (d-a)w - b$ have a <u>double</u> root $\hat{w}$. (That is, let $A = D$.) Then

(i) w_n is infinite for at most one value of n,

(ii) if $\epsilon_0 = -A/mC$ for some positive integer m, w_m is infinite and w_n is finite, $n \neq m$,

(iii) if $\epsilon_0 \neq -A/mC$ for any positive integer m, w_n is finite for all n, and

(iv)
$$w_n = \hat{w}\,[1 + \frac{A\epsilon_0}{A + nC\epsilon_0}] \quad .$$

It turns out that the characteristic polynomial of the difference equation (4(iii)) does indeed have a double root, and moreover it is

$$\hat{w} = \frac{1 + \frac{1}{2}\bar{h}}{1 - \frac{1}{2}\bar{h}} . \qquad \text{Also,} \quad A = \frac{1}{4}\bar{h}^2 , \quad C = 1 - \frac{1}{4}\bar{h}^2$$

<u>Case (i)</u> Assume that there exists no positive integer m such that $\epsilon_0 = -A/mC$. Then w_n is always finite. Let h be fixed, with Re h < 0; then A, C, ϵ_0, and $\hat{w}$ are all fixed. It follows that

$$|\hat{w}| \leq K < 1,$$

and, since

$$1 + \frac{A\epsilon_0}{A + nC\epsilon_0} \to 1 \quad \text{as} \quad n \to \infty ,$$

that there exists a positive integer N such that for all n > N,

$$\left|1 + \frac{A\epsilon_0}{A + nC\epsilon_0}\right| < \frac{K}{|\hat{w}|}$$

It follows from Corollary 2 that

$$\left|\frac{y_{N+n}}{y_N}\right| = \left|\prod_{i=N}^{N+n-1} w_i\right| \leq \prod_{i=N}^{N+n-1} |w_i| < K^{n-1}$$

and hence that $y_{N+n} \to 0$ as $n \to \infty$, since y_N is necessarily finite.

<u>Case (ii)</u> Assume that there exists a positive integer m such that $\epsilon_0 = -A/mC$. Then, in view of Corollary 2, the argument for Case (i) holds if we replace N by $\tilde{N}$, where $\tilde{N} = \max(N, m+1)$.

We have thus established that for all fixed $\bar{h}$, such that Re $\bar{h}$ < 0, the solution of the difference equation resulting from applying method (4) to the test equation $y' = \lambda y$ tends to zero as $n \to \infty$; that is, method (4) is A-stable.

<u>Remark 1</u>. As $\bar{h} \to -\infty$, then $\hat{w} \to 1$ and $w_n \to 1 + \frac{\epsilon_0}{1 - n\epsilon_0}$. Thus method (4) is <u>not</u> L-stable.

<u>Remark 2</u>. In view of the comment concerning method (2) made after the statement of Lemma 1, it is of interest to note that if, for method (4), we also take the additional starting value y_1 to be given by the trapezoidal rule, then once again we

find that $y_{n+1}/y_n = (1+\frac{1}{2}\bar{h})/(1-\frac{1}{2}\bar{h})$, whence A-stability. However, this is not a good choice for the additional starting value, since the trapezoidal rule has order only 2, whereas method (4) has order 3 in general.

Remark 3. It is possible to construct a two-step nonlinear method involving only y_n, y_{n+1}, y_{n+2}, f_n, and f_{n+1}, which, when applied to the test equation $y' = \lambda y$, yields precisely equation (4(iii)), and is therefore A-stable. This method, however, turns out to have order only 2 in general.

Finally, a one-step nonlinear method based on the interpolant $I(x) = (Ax+B)/(x^2+Cx+D)$ can be obtained by requiring that

$$y_{n+j} = I(x_{n+j}),\quad j = 0,1,\quad f_n = I^{(1)}(x_n),\quad f_n^{(1)} = I^{(2)}(x_n),\quad f_n^{(2)} = I^{(3)}(x_n).$$

It is

$$y_{n+1} = y_n + hf_n + \tfrac{1}{2}h^2 f_n^{(1)} + \tfrac{1}{6}h^3 \frac{P_n f_n^{(2)} + hR_n f_n^{(1)}}{P_n + hQ_n - \frac{1}{3}h^2 R_n}, \quad \begin{matrix} |y(x)|+|y^{(1)}(x)|+|y^{(2)}(x)| \neq 0, \\ |y^{(1)}(x)|+|y^{(2)}(x)|+|y^{(3)}(x)| \neq 0, \end{matrix} \tag{5}$$

where

$$P_n = y_n f_n^{(1)} - 2f_n^2,\quad Q_n = f_n f_n^{(1)} - \tfrac{1}{3} y_n f_n^{(2)},\quad R_n = \tfrac{3}{2} f_n^{(1)2} - f_n f_n^{(2)}.$$

Again, the local truncation error is too complicated to quote here, but the order of method (5) turns out to be exactly the same as that of method (4), as given by (4(ii)).

Stability. Applying method (5) to the test equation $y' = \lambda y$ yields

$$\frac{y_{n+1}}{y_n} = \frac{1 + \frac{1}{3}\bar{h}}{1 - \frac{2}{3}\bar{h} + \frac{1}{6}\bar{h}^2},$$

the (1,2) Padé approximation to $e^{\bar{h}}$. The method is therefore A-stable, and is indeed L-stable.

IV. NUMERICAL RESULTS

Linear problem (Fowler and Warten [3])

$$\begin{aligned} y_1' &= -2000y_1 + 1000y_2 + 1 & y_1(0) &= 0 \\ y_2' &= y_1 - y_2 & y_2(0) &= 0 \end{aligned} \qquad 0 \leq x \leq 5$$

Eigenvalues of Jacobian are -2000.5 and -0.5.

Theoretical solution:

$$y_1(x) = -4.97 \times 10^{-4}\exp(-2000.5x) - 5.034 \times 10^{-4}\exp(-0.5x) + .001$$

$$y_2(x) = 2.5 \times 10^{-7}\exp(-2000.5x) - 1.007 \times 10^{-3}\exp(-0.5x) + .001$$

<u>Method (4)</u>

<u>h = 0.1</u>

x	$y_1 \times 10^4$		$y_2 \times 10^4$	
	Numerical	Theoretical	Numerical	Theoretical
0.5	6.5742	6.0776	.2946	2.1551
1.0	1.9358	6.9437	.4663	3.8873
1.5	-8.3263	7.6185	-.5001	5.2371
2.0	9.7032	8.1444	-.0831	6.2888
2.5	.4585	8.5541	.0037	7.1083
3.0	9.2196	8.8734	.0006	7.7468
3.5	1.2109	9.1222	-.0053	8.2443
4.0	8.2871	9.3160	-.0002	8.6320
4.5	2.2690	9.4670	.0034	8.9341
5.0	7.1605	9.5847	.0000	9.1694

<u>h = 0.01</u>

0.5	6.0731	6.0776	2.1467	2.1551
1.0	6.9427	6.9437	3.8854	3.8873
1.5	7.6175	7.6185	5.2371	5.2371
2.0	8.1444	8.1444	6.2901	6.2888
2.5	8.5555	8.5541	7.1106	7.1083
3.0	8.8741	8.8734	7.7492	7.7468
3.5	9.1263	9.1222	8.2592	8.2443
4.0	9.3238	9.3160	8.6525	8.6320
4.5	9.4765	9.4670	8.9564	8.9341
5.0	9.5945	9.5847	9.1915	9.1694

<u>Remarks</u>

1) The first half of the above table displays what would appear to be very inaccurate results for a third-order method with steplength of 0.1. However, inspection of the given system shows that $y_2(0) = y_2^{(1)}(0) = 0$, which is precisely the condition for

method (4) to have order just 1 at the initial point (see(4(iii))). For a first-order method, a steplength of 0.1 is inadequate for this problem. As the second half of the table indicates, reduction of the step-length to 0.01 produces an acceptable solution, which clearly succeeds in reaching the correct steady-state solution, $y_1(x) = y_2(x) = .001$.

2) The numerical solutions with both steplengths indicate that the method has adequate stability. The very large truncation errors introduced initially when the steplength is 0.1 are nevertheless damped as the solution progresses. Note that our theoretical results on stability apply only to systems of the form $\underset{\sim}{y}' = \Lambda\underset{\sim}{y}$, where Λ is a diagonal matrix. For this example, despite the fact that the system is far from diagonal, the stability properties appear to be retained.

3) A comparable explicit linear multistep method would have an interval of absolute stability of roughly (-2, 0). Since there is an eigenvalue of -2000.5, the steplength would need to be less than 0.001 if such a method were to compute a stable solution.

<u>Nonlinear problem (Liniger and Willoughby [7])</u>

$$y_1' = 0.01 - [1 + (y_1 + 1000)(y_1 + 1)][0.01 + y_1 + y_2], \qquad y_1(0) = 0$$

$$y_2' = 0.01 - (1 + y_2^2)(0.01 + y_1 + y_2), \qquad y_2(0) = 0$$

$$0 \leq x \leq 100.$$

Eigenvalues of Jacobian are -1012 and -0.01 at x = 0, and -21.7 and -0.089 at x = 100.

<u>Method (5)</u>. As in the previous example, $y_2(0) = y_2^{(1)}(0) = 0$, so that the method is initially of order only one. If, however, the solution is started at some point to the right of the origin, local reduction of order is avoided. A "theoretical" solution was calculated using a fourth-order RK method with steplength 5×10^{-4}. Using the values given by this solution at x = 1, a solution was calculated using method (5) with steplengths 0.1 and 0.01, giving the following results, at x = 100:

	Theoretical	Numerical, h=0.1	Numerical, h=0.01
y_1	-0.9916	-0.9990	-0.9989
y_2	+0.9833	+0.9940	+0.9939

REFERENCES

[1] Dahlquist, G.G., A special stability problem for linear multistep methods, BIT 3, 27-43 (1963).

[2] Ehle, B.L., On Padé approximations to the exponential function and A-stable methods for the numerical solution of initial value problems, University of Waterloo, Dept. Applied Analysis and Computer Science, Research Report No. CSRR 2010 (1969).

[3] Fowler, M.E. and Warten, R.M., A numerical integration technique for ordinary differential equations with widely separated eigenvalues, IBM Jour., 11, 537-543, 1967.

[4] Lambert, J.D. and Shaw, B., On the numerical solution of y' = f(x,y) by a class of formulae based on rational approximation, Math. Comp. 19, 456-462, 1965.

[5] Lambert, J.D. and Shaw, B., A method for the numerical solution of y' = f(x,y) based on a self-adjusting non-polynomial interpolant, Math. Comp., 20, 11-20, 1966.

[6] Lambert, J.D. and Shaw, B., A generalisation of multistep methods for ordinary differential equations, Numer. Math., 8, 250-263, 1966.

[7] Liniger, W. and Willoughby, R.A., Efficient numerical integration methods for stiff systems of differential equations, IBM Research Report RC-1970, 1967.

[8] Widlund, O.B., A note on unconditionally stable linear multistep methods, BIT, 7, 65-70, 1967.

Curved Elements in the Finite Element Method

A. R. Mitchell and R. McLeod

1. Introduction. One of the most important tasks in the Finite Element Method is the division of the region into suitable non-overlapping elements. This generation of the mesh is particularly important for regions in two and three dimensions. Automatic mesh generation is obviously an important problem for the future, but for the present we shall concentrate on the theoretical background of some of the methods currently in use for dividing up regions in two and three dimensional space enclosed by algebraic curves and surfaces respectively, and show how basis functions can be constructed which take account of the curvature of the external boundary.

2. Two dimensional elements. The only two dimensional elements worth considering are the curvilinear triangle and quadrilateral. These are illustrated in Fig.1. Provided the transformation co-ordinates p and q can be found, every triangular or

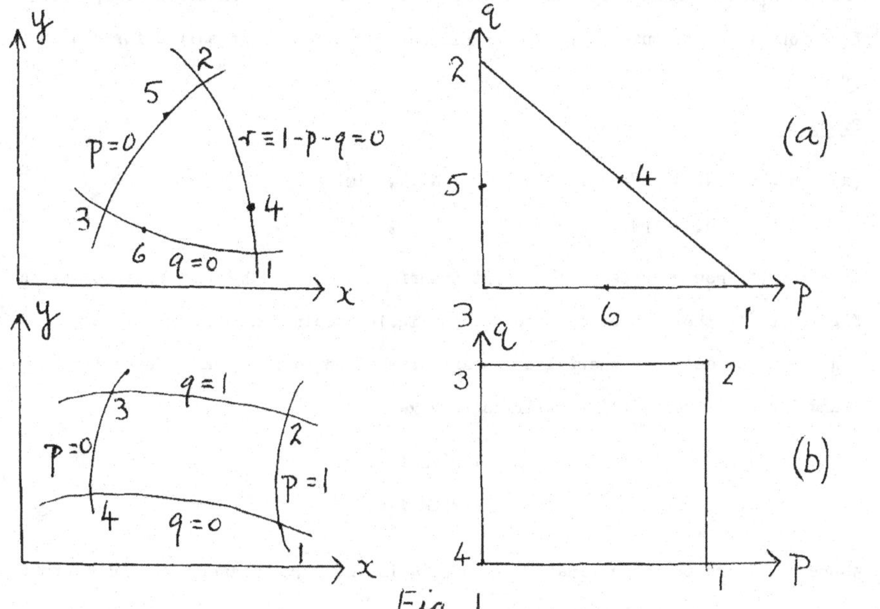

Fig. 1

quadrilateral element can be transformed into the standard triangle or square respectively, and provided the Jacobian of the transformation is well behaved, there is no great problem. In the case of straight sided triangles and parallelograms the transformation formulae are linear, but in all other cases, including the straight sided quadrilateral, the transformation formulae are non-linear and usually difficult to find.

The Triangle

Point transformations.

(1) Lagrange case. Here the transformation formulae are written in the form

$$x = \sum_{i=1}^{n} N_i(p,q)x_i \qquad \left(\sum_{i=1}^{n} N_i(p,q) = 1\right) \qquad (1)$$
$$y = \sum_{i=1}^{n} N_i(p,q)y_i$$

and the $N_i(p,q)$ chosen so that suitable points in the two planes correspond. Equation (1) constitutes a pair of nonlinear equations to be solved for p,q in terms of x,y.

Examples.

(a) $n = 6$ $N_1 = p(2p-1)$. N_2,N_3 similarly

$N_4 = 4pq$. N_5,N_6 " .

Equation (1) now represents the point transformation for the six points shown in Fig.1(a). In the case of 253 and 361 in Fig.1(a) being the straight lines $l = 0$ and $m = 0$ respectively where l and m are normalised to take the value unity at points 1 and 2 respectively, (1) simplifies to give

$$l = p + 2(2L-1)pq$$
$$m = q + 2(2M-1)pq \qquad (1a)$$

where (L,M) are the co-ordinates of point 4 in the (l,m) plane. The transformation co-ordinates p and q isolate to give

$$p^2 + [\frac{2L-1}{2M-1} m - 1 + \frac{1}{2(2M-1)}]p - \frac{1}{2(2M-1)} = 0$$

and

$$q^2 + [\frac{2M-1}{2L-1} l - m + \frac{1}{2(2L-1)}]q - \frac{m}{2(2L-1)} = 0$$

respectively, and the equation of the curved side 142 is the parabola given by

$$(l-m)^2 + \frac{l+m}{2R-1} - \frac{2R}{2R-1} = 0,$$

when $L = M = R$.

(b) $n = 10.$ $N_1 = \frac{1}{2}p(3p-1)(3p-2)$ N_2, N_3 similarly

$N_4 = \frac{9}{2}pq(3p-1)$ N_6, N_8 "

$N_5 = \frac{9}{2}pq(3q-1)$ N_7, N_9 "

$N_{10} = 27pqr$.

This time equation (1) represents the point transformation for ten points, comprising the three vertices, two side points on each side, and one point inside the triangle. In the case of two straight sides and one curved side, the transformation simplifies to give

$$\begin{aligned} l &= p + \frac{9}{2}(6L_{10}-L_4-L_5-1)pq + \frac{27}{2}(L_4-2L_{10})p^2q + \frac{27}{2}(L_5-2L_{10}+\frac{1}{3})pq^2 \\ m &= q + \frac{9}{2}(6M_{10}-M_4-M_5-1)pq + \frac{27}{2}(M_4-2M_{10}+\frac{1}{3})p^2q + \frac{27}{2}(M_5-2M_{10})pq^2, \end{aligned} \quad (1b)$$

where (L_4,M_4), (L_5,M_5), and (L_{10},M_{10}) are the co-ordinates in the (l,m) plane of the two points on the curved side and the point inside the triangle respectively. This time the equation of the curved side is a cubic in l and m except for particular positions of points 4 and 5.

The Lagrange interpolation formula for a general function of l and m, say $U(l,m)$, is

$$U(l,m) = \sum_{i=1}^{n} N_i'(p,q)U_i, \qquad (\sum_{i=1}^{n} N_i'(p,q) = 1) \qquad (2)$$

where p and q are given in terms of l and m by (1). If

$$N_i'(p,q) = N_i(p,q), \qquad i = 1,2,---,n$$

the element is said to be isoparametric, and the basis functions $N_i(p,q)$ $(i=1,2,---,n)$ given by (2) exactly interpolate a linear function of l and m (see (1)), as well as a constant.

(2) Hermite case. This time we go straight to the isoparametric case and consider a triangle with two straight sides and one curved side. Formulae (2) and (1) are replaced by

$$U(l,m) = \sum_{i=1}^{n} N_i(p,q)U_i + [P_j(p,q)(\frac{\partial U}{\partial p})_j + Q_j(p,q)(\frac{\partial U}{\partial q})_j] \tag{3}$$

and

$$l = \sum_{i=1}^{n} N_i(p,q)l_i + \sum_{j=1}^{m}[P_j(p,q)(\frac{\partial l}{\partial p})_j + Q_j(p,q)(\frac{\partial l}{\partial q})_j]$$

$$m = \sum_{i=1}^{n} N_i(p,q)m_i + \sum_{j=1}^{m}[P_j(p,q)(\frac{\partial m}{\partial p})_j + Q_j(p,q)(\frac{\partial m}{\partial q})_j] \tag{4}$$

respectively, where $\sum_{i=1}^{n} N_i(p,q) = 1$, and we are considering only up to first derivatives. The transformation formulae (4) are very complicated, and will be studied in a particular case.

Example $(n = 4, m = 3)$

$N_1 = p(3p - 2p^2 - 7qr)$ $\qquad$ N_2, N_3 similarly

$N_4 = 27pqr$

$P_1 = p(2p^2 - 3p + 1 - r^2 - q^2)$ $\qquad$ $Q_1 = pq(p-r)$

$P_2 = pq(q-r)$ $\qquad$ $Q_2 = q(2q^2 - 3q + 1 - r^2 - p^2)$

$P_3 = rp(1 - p - 2q)$ $\qquad$ $Q_3 = rq(1 - 2p - q)$

Equation (4) now gives the point transformation formulae for the case of the function and its two first order partial derivatives at each of the three vertices together with the function at an internal point. The transformation formulae can be simplified to give

$$\begin{aligned} l = {} & p(3p-2p^2) + 27(L_{10}-7)pqr \\ & + p(p^2-p+2qr)\left(\frac{\partial l}{\partial p}\right)_1 + pq(p-r)\left(\frac{\partial l}{\partial q}\right)_1 \\ & + pq(q-r)\left(\frac{\partial l}{\partial p}\right)_2 + q(q^2-q+2pr)\left(\frac{\partial l}{\partial q}\right)_2 \\ & + pr(r-q)\left(\frac{\partial l}{\partial p}\right)_3 + qr(r-p)\left(\frac{\partial l}{\partial q}\right)_3 \end{aligned} \tag{5}$$

$$m = \text{similarly}$$

where (L_{10}, M_{10}) are the co-ordinates of the internal point in the (l,m) plane. Two possible solutions of (5) are

$$\begin{aligned} l &= bp + [-\tfrac{1}{2} - 4b + \tfrac{27}{2} L_{10}]p^2 + 3[\tfrac{1}{2} + b - \tfrac{9}{2} L_{10}]p^3 \\ m &= cq + [-\tfrac{1}{2} - 4c + \tfrac{27}{2} M_{10}]q^2 + 3[\tfrac{1}{2} + c - \tfrac{9}{2} M_{10}]q^3 \end{aligned} \tag{6}$$

where b and c are arbitrary constants, and

$$\begin{aligned} l &= p - [\tfrac{1}{3}(d+e) - 9L_{10} + 3]pq + dp^2q + epq^2 \\ m &= q - [\tfrac{1}{3}(f+g) - 9M_{10} + 3]pq + fp^2q + gpq^2 , \end{aligned} \tag{7}$$

where d,e,f, and g are arbitrary constants. In fact (7) is identical with (1b). A variety of boundary curves can be obtained by eliminating p and q from either (6) or (7) for different values of the arbitrary constants and for different locations of the point (L_{10}, M_{10}).

Direct Transformations.

So far the point transformations considered have been of the form

$$\begin{aligned} l &= l(p,q) \\ m &= m(p,q) \end{aligned} \tag{8}$$

with the curved side given by

$$1 - p - q = 0, \tag{9}$$

and the integrals to be evaluated of the form

$$\iint_\Delta F(l,m)\,dl\,dm = \iint_\Delta F(p,q)J\begin{pmatrix} l & m \\ p & q \end{pmatrix} dpdq = \iint_\Delta G(p,q)dpdq \tag{10}$$

Provided the Jacobian J is well behaved for all points (l,m) in the curved triangle, the evaluation of the integral in (10) presents no problem. The equation of the curved side, given by (9), is difficult to obtain in closed form, however, and in general the best that can be obtained is an approximate solution of (8) for p and q. Even if an exact solution of (8) were possible, it is almost certain that the curve obtained from (9) will not coincide with the given curve unless at the points selected by the transformation formula.

To obviate this serious defect of point transformations of the form (8), we now look at the direct transformations

$$\begin{aligned} p &= p(l,m) \\ q &= q(l,m) \end{aligned} \tag{11}$$

which transform the line $l = 0$ onto $p = 0$, the line $m = 0$ onto $q = 0$, and the curve $f(l,m) = 0$ onto $1 - p - q = 0$. Also $(0,m)$ and $(l,0)$ are invariant under transformation. An example of such a transformation is

$$\begin{aligned} p &= \frac{-l(al+bm-1)}{1 - al - cm} \\ q &= \frac{-m(cm+bl-1)}{1 - al - cm} \quad , \end{aligned} \tag{12}$$

where $1 - al - cm > 0 \ \forall \ l,m$ in the element, and the curve is given by

$$f(l,m) \equiv al^2 + 2blm + cm^2 - (1+a)l-(1+c)m + 1 = 0. \tag{13}$$

If we use the basis functions

$$\begin{aligned} N_1 &= p(2p-1) \qquad && N_2,N_3 \text{ similarly} \\ N_4 &= 4pq && N_5,N_6 \text{ similarly,} \end{aligned} \tag{14}$$

the integrals to be evaluated have the form

$$\iint_\Delta F(l,m)dldm = \iint_\Delta \frac{F(p,q)}{J\begin{pmatrix} p & q \\ l & m \end{pmatrix}} \quad dpdq \ ,$$

and this presents no great problem provided numerical integration is used. This time the basis functions obtained by substituting the values of p and q obtained from (12) in (14) do <u>not</u> exactly interpolate a linear function of l and m, not even in the special case when

$$a = c = -b = -\frac{2R-1}{2R} \ ,$$

and (13) coincides with (1a).

Construction in the Physical Plane.

We now use ideas from three dimensional geometry to construct basis functions for the triangle with two straight sides and one curved side in the physical plane.

Consider the family of surfaces $z(l,m) = 0$ which intersects the (l,m) plane in the curve $f(l,m) = 0$, and is given by the equation

$$z(\alpha z + \beta l + \gamma m + \delta) + f(l,m) = 0, \qquad (15)$$

where α, β, γ and δ are parameters. If the curve is taken to be the general conic

$$f(l,m) \equiv al^2 + blm + cm^2 - (1+a)l - (1+b)m + 1 = 0$$

which passes through the points (1,0) and (0,1), and is normalised at (0,0), and if we impose the conditions

$$z = 1 - l \quad \text{at} \quad m = 0$$

$$z = 1 - m \quad \text{at} \quad l = 0$$

on the surface, then (15) becomes

$$\alpha z^2 + [\alpha(l+m-1)+(al+cm-1)]z + [al^2+blm+cm^2-(1+a)l - (1+b)m+1] = 0 \qquad (16)$$

Any surface $z(l,m) = 0$ which satisfies (16) represents a basis function which takes the value unity at the origin, is linear along the l and m axes, and is zero on the general conic.

The part basis function (wedge) which is a solution of (16) is $W_3(l,m)$, and the remaining wedges $W_i(i = 1,2,4)$ associated with the points $i(i = 1,2,4)$ are obtained from the equations

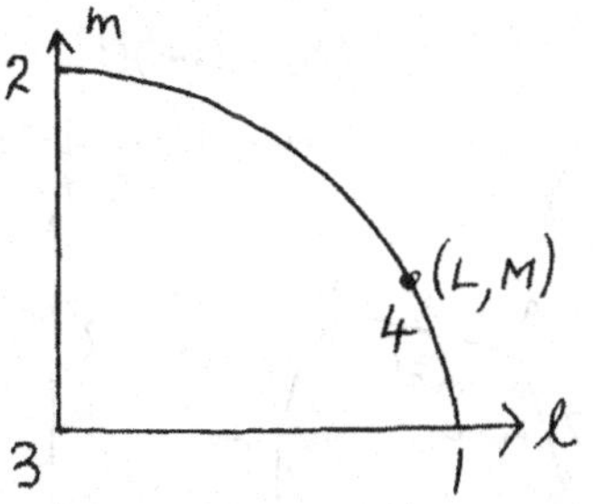

$$W_1 + W_2 + W_4 = 1 - W_3$$

$$W_1 + W_4 L = l \qquad (17)$$

$$W_2 + W_4 M = m$$

The relations (17) ensure that the wedges form a basis for linear approximation over the triangular element. Equations (17) solve to give

$$W_1 = \frac{(1-M)l + Lm - L}{1 - L - M} + \frac{L}{1 - L - M} W_3$$

$$W_2 = \frac{Ml + (1-L)m - M}{1 - L - M} + \frac{M}{1 - L - M} W_3 \qquad (18)$$

$$W_4 = \frac{1 - l - m}{1 - L - M} - \frac{1}{1 - L - M} W_3$$

Special cases of (16) worthy of special mention are

(i) $\alpha = 0$. This leads to the rational wedge functions of Wachspress if $a,c \neq 0$. When $a = c = 0$, $W_3 = 1 - l - m + blm$.

(ii) $\alpha = \frac{2R-1}{2R}$, $a = -b = c = -\frac{2R-1}{2R}$. This time (16) corresponds to the simplest isoparametric case and z represents $1 - p - q$ where p and q are given by (1a).

The cases where the wedge functions are quadratic and cubic instead of linear along the straight sides require extra nodal points and are dealt with by the present authors in an earlier paper.

THE QUADRILATERAL

Following the study for the triangle, we start with the point transformations given by (1), where this time p and q are as illustrated in Fig. 1(b). If we choose the example of the quadrilateral with one intermediate point on each side, the coefficients in (1) are given by

$$\begin{aligned} N_1 &= pq(2p+2q-3) & N_5 &= 4pq(1-p) \\ N_2 &= (1-p)q(-2p+2q-1) & N_6 &= 4q(1-p)(1-q) \\ N_3 &= (1-p)(1-q)(1-2p-2q) & N_7 &= 4p(1-p)(1-q) \\ N_4 &= p(1-q)(2p-2q-1) & N_8 &= 4pq(1-q) \; . \end{aligned} \qquad (19)$$

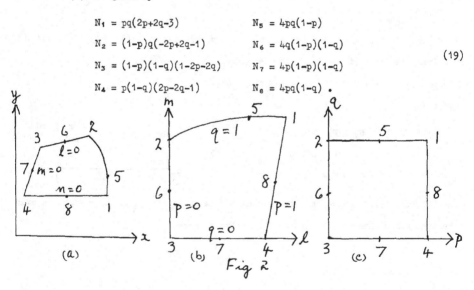

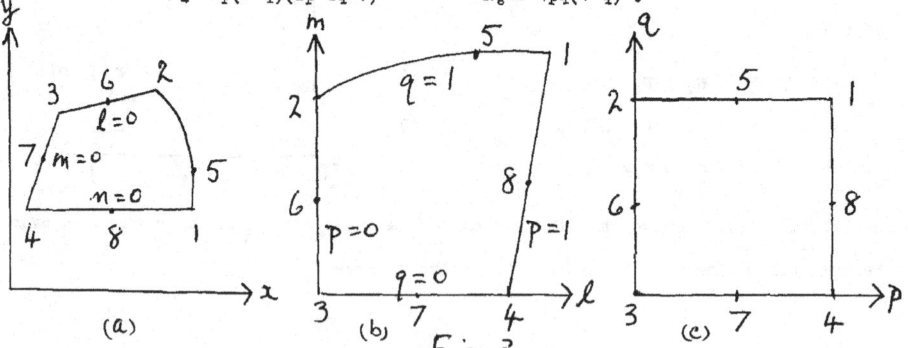

Fig 2

For the case of three straight sides and one curved side (Fig. 2), the transformation formulae (1) with the coefficients given by (19) reduce to

$$\begin{aligned} l &= p + Apq + Bp^2q \\ m &= q + Cpq + Dp^2q \ , \end{aligned} \tag{20}$$

where

$$A = -2 - \alpha + 4L \qquad B = 2 + 2\alpha - 4L$$

$$C = -3 - \beta + 4M \qquad D = 2 + 2\beta - 4M.$$

with $4 \equiv (1,0)$, $2 \equiv (0,1)$, $1 \equiv (1+\alpha,\beta)$, and $5 \equiv (L,M)$. The curve $q = 1$ is the parabola

$$[Dl + B(1-m)]^2 + (D+AD-BC)[Cl + (1+A)(1-m)] = 0 \ .$$

In the important case, $\alpha = 0$, $L = \frac{1}{2}$, (20) can be solved to give

$$\begin{aligned} p &= l \\ q &= \frac{m}{1 + Cl + Dl^2} \end{aligned} \tag{21}$$

Otherwise the solution of (20) for p and q is not a trivial matter. If the quadrilateral element is isoparametric, the basis functions (19), with p and q given in terms of l and m by (20), interpolate any linear function of l and m.

In order to cope with more general boundary shapes, we can consider the direct transformations

$$p = p(l,m)$$
$$q = q(l,m)$$

for the quadrilateral with $\alpha = 0$, $L = \frac{1}{2}$. These are required to transform the line $l = 0$ onto $p = 0$, the line $m = 0$ onto $q = 0$, the line $l = 1$ onto $p = 1$, and the curve $f(l,m) = 0$ onto $q = 1$. In addition, $(0,m)$, $(1,\frac{m}{\beta})$, and $(1,0)$ are invariant under transformation.

Finally for the quadrilateral we consider construction of the basis functions in the physical plane. For the case of three straight sides and one curved side (Figs. 2(a) and 2(b)), the quadrilateral taken to be a "triangle" made up of two straight sides 34 and 32, and a curved side 214. If the true curve 12 is second order, then the "curve" 214 will be cubic. This cubic curve passes through (0,1) and (1,0) and its general form has the equation

$$F(l,m) \equiv (al^2 + blm + cm^2 + dl - (1+c)m + 1)(-l + em + 1) = 0,$$

with

$$d = \frac{(1+c)\beta - 1 - a(1+\alpha)^2 - b(1+\alpha)\beta - c\beta^2}{1 + \alpha}$$

and

$$e = \frac{\alpha}{\beta}.$$

In a similar manner to that adopted for the triangle, the surface $z(l,m) = 0$ which is linear along 34 and 32 and passes through $F(l,m) = 0$ when $z = 0$ satisfies the equation

$$z[azl - eczm - (a+d)l + (ec-e+c)m - 1] + F(l,m) = 0 \qquad (22)$$

Any solution of (22) will be a suitable basis function W_3 to be associated with node 3. Repeating the process with the "triangle" composed of the two straight sides 41 and 43, and a curved side 321 a basis function W_4 is obtained for node 4. The remaining basis functions for the nodes 1, 2, and 5 are obtained from the relations

$$\sum_{i=1}^{5} W_i = 1, \qquad \sum_{i=1}^{5} l_i W_i = l, \qquad \sum_{i=1}^{5} m_i W_i = m.$$

Full details of the derivation of basis functions which are (i) linear, (ii) quadratic on the straight sides of a quadrilateral with three straight sides can be found in the Ph.D. thesis of R. McLeod.

3. Transformation of the complete region. In section 2, it is assumed that the region in the physical plane is already divided up into elements and transformation and direct methods described for dealing with the elements individually. This way the interior elements are independent of the boundary shape, whilst the elements round the edge each include a small part of the boundary. In recent papers, Gordon and Hall, and Zienkiewicz and Phillips propose to map the complete region in the problem domain onto a unit square in the transformed domain. The latter is then

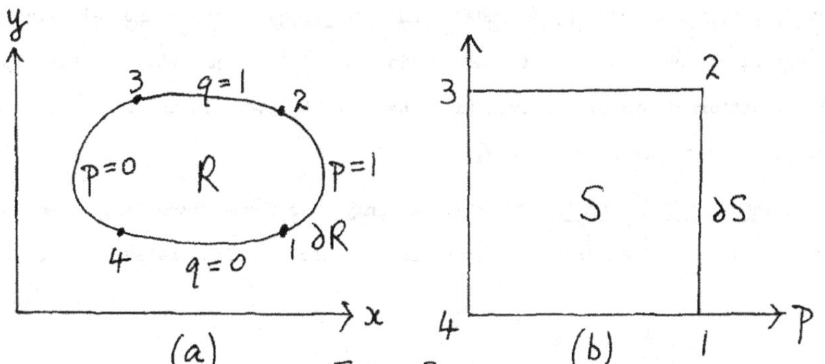

Fig. 3

subdivided by lines parallel to the p and q axes and the grid formed transformed back to the region in the physical plane. This method of generating elements in the problem domain is trivial for regions such as circles and ellipses in E^2 and circular cylinders and spheres in E^3. For regions where curvilinear coordinate systems do not exist, the following procedure is advocated.

1. Consider the mapping $\underline{F} : S \to R$ where $S : [0,1] \times [0,1]$, and $\underline{F}$ is given by

$$\underline{F}(p,q) = \begin{bmatrix} x(p,q) \\ y(p,q) \end{bmatrix} .$$

2. Select four points on ∂R, and identify these as corresponding to the four corners of S in order. i.e. these are the four points with coordinates $\underline{F}(1,0)$, $\underline{F}(1,1)$, $\underline{F}(0,1)$, and $\underline{F}(0,0)$ respectively.

3. These four points separate ∂R into four segments which are the curves $\underline{F}(1,q)$, $\underline{F}(p,1)$, $\underline{F}(0,q)$, and $\underline{F}(p,0)$ respectively.

4. (Gordon and Hall only) Define a bilinearly blended transfinite map $\underline{I}(p,q)$ given by

$$\underline{I}(p,q) = (1-p)\underline{F}(0,q) + p\underline{F}(1,q) + (1-q)\underline{F}(p,0) + q\underline{F}(p,1) - (1-p)(1-q)\underline{F}(0,0) - (1-p)q\underline{F}(0,1) - p(1-q)\underline{F}(1,0) - pq\underline{F}(1,1),$$

where $\underline{I} = \underline{F}$ for points (p,q) on the perimeter of S.

The above procedure, of course, depends on knowing $\underline{F}(p,q)$. Gordon and Hall give no general method for obtaining the transformation formula. Zienkiewicz and Phillips use point transformations of the type (i), where the $N_i(p,q)$, in the simplest case of parabolic curves, are given by (19).

4. Three dimensional elements. The most commonly used three dimensional elements are the curvilinear tetrahedron and hexahedron. These are illustrated in Fig. 4.

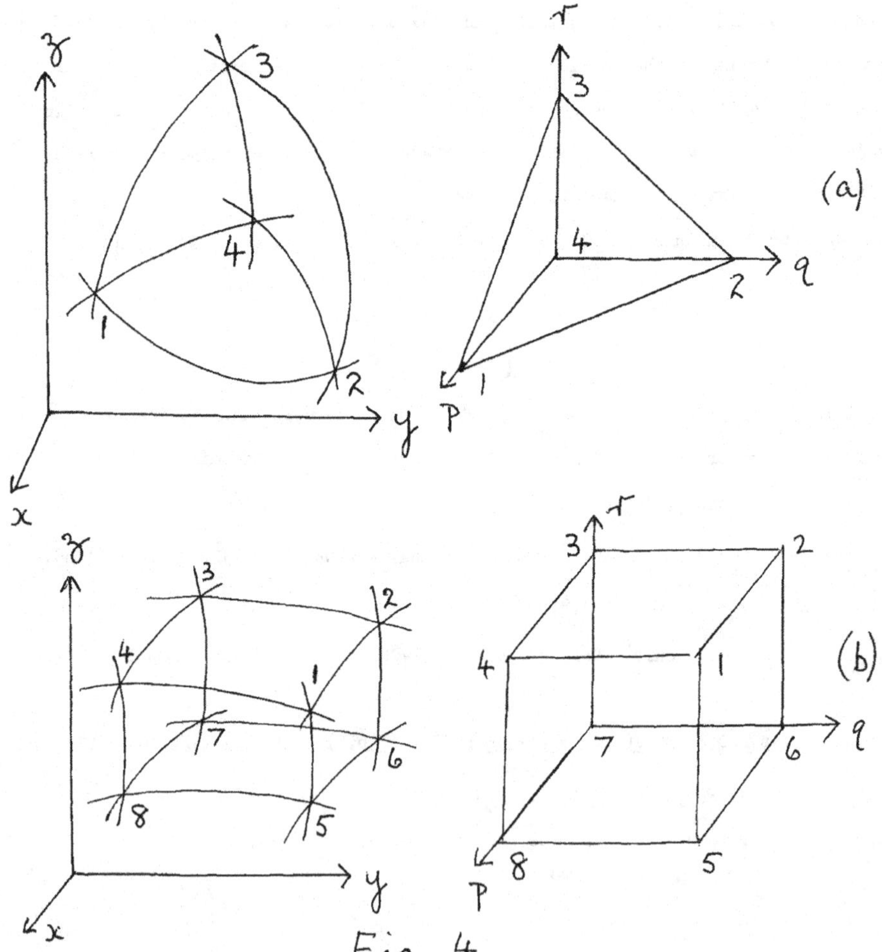

Fig. 4.

Point Transformations.

The transformation formulae are written in the form

$$x = \sum_{i=1}^{n} N_i(p,q,r)x_i$$

$$y = \sum_{i=1}^{n} N_i(p,q,r)y_i \qquad \left(\sum_{i=1}^{n} N_i(p,q,r) = 1 \right) \tag{23}$$

$$z = \sum_{i=1}^{n} N_i(p,q,r)z_i$$

and the $N_i(p,q,r)$ chosen so that suitable points in the two regions correspond.

Examples.

(a) Tetrahedron $n = 10$.

$N_1 = p(2p-1)$ at 1. 2,3,4 similarly.

$N_2 = 4pq$ at mid point of 12. other mid points similarly.

(b) Hexahedron $n = 20$.

$N_1 = pqr(2p+2q+2r-5)$ at 1. 2,3,...,8 similarly

$N_9 = 4pqr(1-p)$ at mid point of 12. other mid points similarly.

In order to obtain the equation of a curved surface, say 123 of the tetrahedron, we eliminate p,q, and r between (23) and the relation

$$1 - p - q - r = 0.$$

This leads to a quartic surface in x,y and z. In a similar manner, the equation of the curved surface 1234 of the hexahedron is obtained by eliminating p,q, and r between (23) and the relation

$$1 - r = 0.$$

This leads to a sextic surface in x,y, and z. It should be pointed out that although point transformations are relatively simple to use for regions in three dimensions, an unsurmountable difficulty appears to be the derivation of the equation of the curved surface which is implicit in the point transformation.

Construction in the Physical Space

The counterpart in three dimensions of the triangle with two straight sides and one curved side is the tetrahedron with three plane faces and one curved face. Unfortunately it is not possible to divide up a finite region in three dimensional space enclosed by a curved surface using non parallel planes so that the elements are plane tetrahedra inside the region and tetrahedra with three plane faces and one curved face round the boundary of the region. The reason for this is that two oblique planes meet in a line and a line cuts the surface of the region in only two points. Consequently all over the surface, elements in the shape of orange segments are duplicated by the two types of tetrahedral element which complete the original region. (See Fig.5). This criticism applies equally to point transformations of the type illustrated in Fig.4(a) when the tetrahedral element in the physical plane has three plane faces and one curved face.

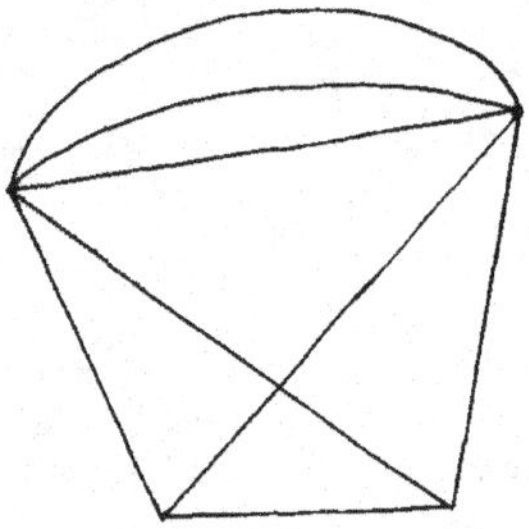

Fig. 5.

An alternative direct method for dealing with the subdivision of a region in three dimensions is to consider the intersection of a block of regular tetrahedra or hexahedra with the region. The elements round the boundary of the region are then of three distinct types in the case of the tetrahedra and of seven distinct types in the case of the hexahedra. One, two and three vertices can be cut off in turn from the regular tetrahedron, and one, two, ---, six and seven vertices in turn in the case of the regular hexahedron. Basis functions can be constructed both for the tetrahedra and the hexahedra, but in the latter case, in particular, the procedure is very complicated.

<u>Example</u>. Construct basis functions for the tetrahedron with three plane faces and one curved face (see Fig.6), the latter being part of a quadric surface with centre at node 1.

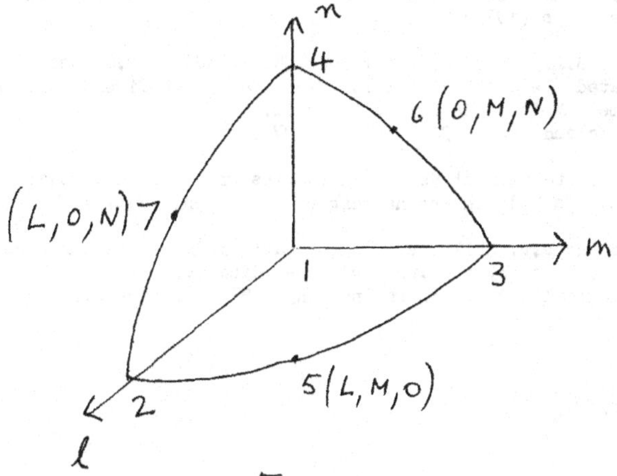

Fig. 6.

The three plane faces are l = 0, m = 0, and n = 0 respectively, where each quantity is normalised at the opposite vertex.

Much remains to be done in order to cope successfully with curved boundaries and interfaces for regions in three dimensions. The authors consider this problem to be of the same order of difficulty and importance as the problems of singularity and nonlinearity, areas which have attracted much larger numbers of research workers.

6. <u>References</u>.

P.G. Ciarlet and P.A. Raviart. The Combined Effect of Curved Boundaries and Numerical Integration in Isoparametric Finite Element Methods. Symposium at Maryland (1972). Academic Press.

W.J. Gordon and C.A. Hall. Geometric Aspects of the Finite Element Method : Construction of Curvilinear Coordinate Systems and their Application to Mesh Generation. General Motors Research Publication No.1286. (1972).

R. McLeod and A.R. Mitchell. The Construction of Basis Functions for Curved Elements in the Finite Element Method. J.I.M.A. (1972), <u>10</u>, 382-393.

R. McLeod. Basis Functions for Curved Elements in the Finite Element Method. Ph.D. Thesis (1972) University of Dundee.

A.R. Mitchell. Curved Elements in the Finite Element Method.
Proc. Second Manitoba Conference on Numerical Mathematics (1972).

E.L. Wachspress. A Rational Basis for Function Approximation Part II. Curved Sides.
J.I.M.A. (1973).

E.L. Wachspress. Algebraic - Geometry Foundations for Finite Element Computation.
These proceedings (1973).

O.C. Zienkiewicz, B.M. Irons, J. Ergatondis, S. Ahmad and F.C. Scott. Isoparametric and Associated Element Families for Two - and Three-Dimensional Analysis.
Finite Element Methods in Stress Analysis.
Editors I. Holand and K. Bell. Tapir 1972.

M. Zlamal. The Finite Element Method in Domains with Curved Boundaries.
Int. J. Num. Methods in Engineering (1973) 5, pp.367-373.

O.C. Zienkiewicz and D.V. Phillips. An automatic mesh generation scheme for plane and curved surfaces by isoparametric co-ordinates.
Int. J. Numerical Methods in Engineering (1971) 3, 519-528.

The Design of Difference Schemes for Studying Physical Instabilities

K.W. Morton

Introduction

One of the most important areas of plasma physics research is the study and elimination of magnetohydrodynamic instabilities. A wide range of equilibria are possible in which the plasma pressure is balanced by magnetic forces - embodied in the equation

$$\underline{\nabla} p = \frac{1}{c}(\underline{J} \times \underline{B}). \qquad (1)$$

These are not always easy to find but an even more difficult task is to determine whether or not they are stable. The initial step is to consider the linearised equations governing magnetohydrodynamic perturbations and see whether there are any growing modes.

Several approaches to this problem are possible, some of which using a variational energy principle yield just a yes/no answer. However, methods which obtain the fastest growing (a least damped) mode and its growth rate are especially useful since this provides some of the best data for comparing theory with experimental observation. Of these methods, the one which is of most general applicability and provides the most information consists of straight forwardly solving the perturbation equations as an initial-boundary value problem with arbitrary initial data.

In this paper we discuss the design of difference methods for this purpose and describe some early results obtained from experiments with model problems. The work has been carried out in collaboration with A. Sykes and J. A. Wesson of UKAEA, Culham Laboratory and details of the calculations will be described elsewhere in joint papers with them: Figs. I, IV-VII are reproduced by kind permission of the Laboratory. Several years ago, experiments were made with one dimensional problems by both J. A. Wesson and J. Killeen - both unpublished. We are concerned here with two dimensional problems.

From a glance at the perturbation equations, which are given in the Appendix, it is clear their complexity is in itself a challenge. In addition, however, one has to be extremely careful that there are no weak numerical instabilities which can be mistaken for real ones nor, on the other hand, excessive damping inhibiting the appearance of the real modes. This calls for careful attention to energy conservation and the proper implementation of boundary conditions.

In some problems, too, the growth rates of interest may be very slow compared with the fastest waves in the system. Thus ideally one would like to devise implicit methods of wide stability range and accurate representation of the important growth rates. Fortunately the linearity of the equations and their special form enables considerable progress to be made and successful explicit schemes have been devised. Whether or not useful implicit schemes can be designed is, however, still an open question.

Model Problems for the Time Differencing

A simple, rather artificial, model problem is provided by the wave equation with a forcing term

$$u_{tt} = c^2 u_{xx} + Au. \qquad (2)$$

Fourier modes $e^{i(kx+\omega t)}$ lead to the dispersion relation

$$\omega^2 = c^2k^2 - A \qquad (3)$$

so that only the larger wavelength modes, with $c^2k^2<A$, yield a growing-damping pair : in this case the fastest growing mode is the largest that the system can maintain.

Simple though this model is, it allows us to look at some of the criteria for accurate time differencing. Consider the general three-level scheme

$$\frac{u^{n+1} - 2u^n + u^{n-1}}{(\Delta t)^2} = \frac{c^2\delta_x^2(\alpha u^{n+1} + \beta u^n + \gamma u^{n-1})}{(\Delta x)^2} + A(du^{n+1} + eu^n + fu^{n-1}) \qquad (4)$$

where $\alpha+\beta+\gamma = 1$, $d+e+f = 1$. It is well known that numerical stability requires that

$$\gamma \leq \alpha, \quad (\beta-\alpha-\gamma)(c\,\frac{\Delta t}{\Delta x})^2 < 1. \qquad (5)$$

A special requirement in the present case is that all wave numbers $k^2 < A/c^2$, and no others, lead to a growing-damping pair of modes. This is ensured if, in addition to (5), we have

$$4(f-\gamma)\; c^2(\frac{\Delta t}{\Delta x})^2 \leq 1, \quad 4(d-\alpha)\; c^2(\frac{\Delta t}{\Delta x})^2 \leq 1, \quad 2(\beta-e)\; c^2(\frac{\Delta t}{\Delta x})^2 \leq 1. \qquad (6)$$

One of these conditions is, of course, redundant and they may be summarised by saying that the forcing term must be centrally weighted at least to the same extent as the wave term.

However, accurate treatment of this stability transition places even more severe constraints on the choice of d, e and f. One wishes to take Δt so that while $(A-c^2k^2)(\Delta t)^2$ is small for the growing modes nevertheless $A(\Delta t)^2$ may be quite large. Now the amplification factors will be functions of

$$(dA - \alpha\frac{4c^2 \sin^2 \frac{1}{2}k\Delta x}{(\Delta x)^2})\;(\Delta t)^2 \quad , \text{etc.} \qquad (7)$$

and poor accuracy will result unless these expressions are small when $(A-c^2k^2)\ (\Delta t)^2$ is small. Hence we are led to the choice

$$d = \alpha,\ e = \beta,\ f = \gamma \qquad (8)$$

which is amply confirmed by numerical experiment. In general, it means that all terms contributing to the calculation of marginal stability should be given the same time level treatment.

The effect of introducing a second space dimension may be considered with the help of the model equation

$$u_{tt} = a\,u_{xx} + 2bu_{xy} + cu_{yy} + Au, \quad b^2 < a\,c. \qquad (9)$$

In a recent paper (3), McKee has surveyed the methods available for such equations and suggested a new A.D.I. scheme. None of the unconditionally stable methods seem to be suitable to the present situation.

Stability is achieved by treating au_{xx} and cu_{yy} implicitly, the equations being made practically soluble by including each alternately in an A.D.I. or fractional step arrangement. In all cases, the cross derivative term is treated explicitly so as to avoid the solution of a wide band matrix. But the considerations described above suggest that this is undesirable. A. Sykes has carried out extensive trials with the conclusion that as soon as the greater range of the implicit methods is exploited the accuracy becomes unacceptable. He has, however, devised schemes with improved stability properties for variable co-efficient problems compared with those given in McKee's paper and all hope of designing an accurate, practical implicit scheme has not been abandoned. For the moment effort has been concentrated on that old faithful, the leap frog method, with the attendant limitation that only problems with relatively fast-growing modes can be tackled.

Space Differencing

The most straightforward differencing of the right hand side of (9) consists of its replacement by

$$(a\delta_x^2 u^n + 2b\,\Delta_{ox}\,\Delta_{oy} u^n + c\delta_y^2\, u^n)/(\Delta x)^2, \qquad (10)$$

in the usual notation $\delta u_j = u_{j+\frac{1}{2}} - u_{j-\frac{1}{2}}$, $\Delta_o\, u_j = \frac{1}{2}(u_{j+1} - u_{j-1})$ and the subscripts denoting the variable to be differenced. However, the variable coefficients quickly generate weakly unstable oscillations which ruin the solution : typical early development of such oscillations is shown in Figure I.

Figure I

Displacement amplitude
for stability calculation
on a rectangle.

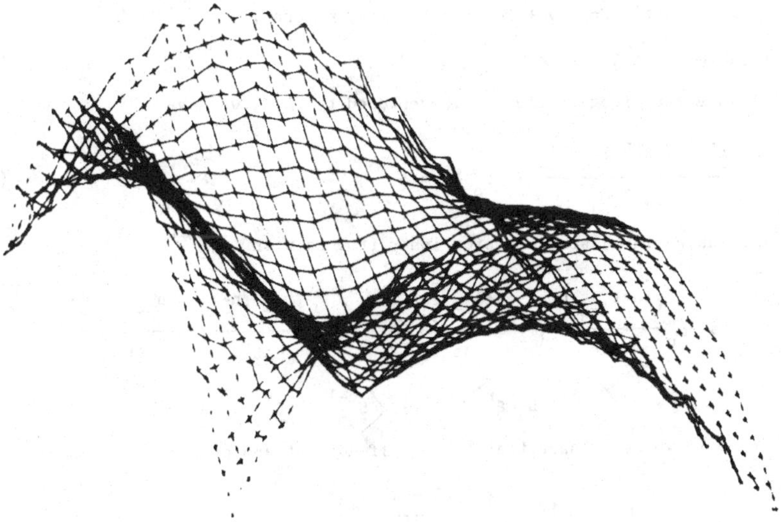

Rather than damping these out in the usual rather arbitrary way with the chance of affecting growth rates, a better spatial differencing is called for with proper attention to energy conservation. As this is a linear problem such conservation should be very effective and should be implemented as precisely as possible. We proceed along the lines described by Morton[4].

The space operator in the perturbation equation (A1) is self-adjoint in a bounded region V under the boundary condition that the normal component of the perturbation displacement is zero on the boundary. Denoting it by - L, the equation takes the form

$$\rho_o \frac{\partial^2 \underline{\xi}}{\partial t^2} + L\,\underline{\xi} = 0. \tag{11}$$

we denote by $\langle\cdot,\cdot\rangle_V$ the inner product over the volume V of two vectors whose scalar product is taken before integration and by $\langle\cdot,\cdot\rangle_S$ the inner product of two scalars.

Then taking the inner product of (11) with $\partial\underline{\xi}/\partial t$ we obtain

$$\left\langle \frac{\partial\underline{\xi}}{\partial t}, \rho_o \frac{\partial^2\underline{\xi}}{\partial t^2}\right\rangle_V + \left\langle \frac{\partial\underline{\xi}}{\partial t}, L\underline{\xi}\right\rangle_V = 0$$

i.e.
$$\frac{d}{\partial t}\left[\tfrac{1}{2}\left\langle \frac{\partial\underline{\xi}}{\partial t}, \rho_o \frac{\partial\underline{\xi}}{\partial t}\right\rangle_V + \tfrac{1}{2}\left\langle \underline{\xi}, L\underline{\xi}\right\rangle_V\right] = 0 \tag{12}$$

i.e. Kinetic Energy + Potential Energy = const.

because
$$\langle \underline{\xi}, L\underline{\eta}\rangle_V = \langle L\underline{\xi}, \underline{\eta}\rangle_V. \tag{13}$$

If in a leap-frog scheme L is replaced by L_Δ, we have

$$\rho_o \frac{\underline{\xi}^{n+1} - 2\underline{\xi}^n + \underline{\xi}^{n-1}}{(\Delta t)^2} + L_\Delta\,\underline{\xi}^n = 0 \tag{14}$$

and taking the discrete vector inner product with $\tfrac{1}{2}(\underline{\xi}^{n+1} - \underline{\xi}^{n-1})$ gives

$$\tfrac{1}{2}\left\langle \rho_o \frac{(\underline{\xi}^{n+1} - \underline{\xi}^n)}{\Delta t}, \frac{\underline{\xi}^{n+1} - \underline{\xi}^n}{\Delta t}\right\rangle_V - \tfrac{1}{2}\left\langle \rho_o \frac{\underline{\xi}^n - \underline{\xi}^{n-1}}{\Delta t}, \frac{\underline{\xi}^n - \underline{\xi}^{n-1}}{\Delta t}\right\rangle_V$$

$$+ \tfrac{1}{2}\left\langle \underline{\xi}^{n+1}, L_\Delta\,\underline{\xi}^n\right\rangle_V - \tfrac{1}{2}\left\langle \underline{\xi}^{n-1}, L_\Delta\underline{\xi}^n\right\rangle_V = 0. \tag{15}$$

Hence if we can ensure that L_Δ is self-adjoint and define

$$\text{K.E.} = \tfrac{1}{2}\left\| \rho_o^{\frac{1}{2}}\left(\frac{\underline{\xi}^{n+1} - \underline{\xi}^n}{\Delta t}\right)\right\|_V^2$$

(16)

$$\text{P.E.} = \tfrac{1}{2}\left\langle \underline{\xi}^{n+1}, L_\Delta\,\underline{\xi}^n\right\rangle_V$$

we will achieve exact energy conservation with the difference scheme from one pair of time levels to the next.

Consideration of expressions (A3) and (A4) in the Appendix show that the self adjointness of L results from two vector identities

$$\underline{\nabla}\cdot(a\underline{u}) = (\underline{u}\cdot\underline{\nabla})a + a\underline{\nabla}\cdot\underline{u} \qquad (17a)$$

and

$$\underline{\nabla}\cdot(\underline{u} \times \underline{v}) = \underline{v}\cdot(\underline{\nabla} \times \underline{u}) - \underline{u}\cdot(\underline{\nabla} \times \underline{v}) \qquad (17b)$$

from which it follows that

$$\left\langle \underline{u}, \underline{\nabla} a \right\rangle_V = -\left\langle \underline{\nabla}\cdot\underline{u}, a \right\rangle_S \qquad \text{if } u_n = o \text{ on } S \qquad (18a)$$

and

$$\left\langle \underline{u}, \underline{\nabla} \times \underline{v} \right\rangle_V = \left\langle \underline{\nabla} \times \underline{u}, \underline{v} \right\rangle_V \qquad \text{if } \underline{u} \times \underline{v}\cdot\hat{\underline{n}} = o \text{ on } S. \qquad (18b)$$

That is, -grad is the adjoint of div and curl is self-adjoint under the appropriate boundary conditions : since $\underline{u}$ is related to $\underline{v}$ in the present use of (18b) we do not need to consider the general conditions for its validity.

To obtain these properties in the difference scheme it is imperative that terms like u_{xx}, u_{xy} are treated in a co-ordinated way. The essential grouping are the vector operators grad, div and curl; these must be treated consistently whenever they occur as well as being properly related to each other. To simplify the situation we will consider first the reduced equation (in two dimensions)

$$\rho_o \frac{\partial^2 \underline{\xi}}{\partial t^2} = \underline{\nabla}(\Gamma\, \underline{\nabla}\cdot\underline{\xi}). \qquad (19)$$

Then, leaving aside boundary conditions, property (18a) may be retained by using the central differences Δ_{ox}, Δ_{oy} everywhere. Unfortunately this does not give a very compact scheme, particularly for terms like u_{xx}, although its virtues were recognized and exploited by Courant, Friedrichs & Lewy (1) in their classical paper on the subject.

A much more compact scheme is provided by putting

$$\underline{\nabla}_\Delta \equiv \left(\frac{\mu_y \delta_x}{\Delta x}, \frac{\mu_x \delta_y}{\Delta x}\right)$$

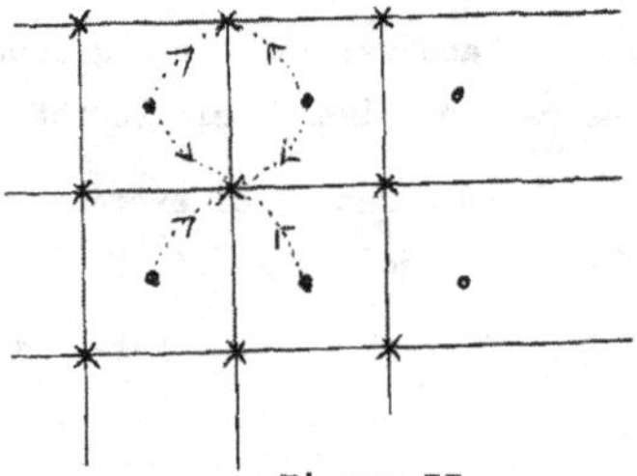

Figure II

on a square, cartesian mesh: thus if $\underline{\xi}$ is defined on the integral points of the mesh, marked with a cross on the diagram, $\underline{\nabla}_\Delta \cdot \underline{\xi}$ is defined at the half integral points, marked with a dot and (19) is replaced by a nine point scheme. Note that it will result in u_{xx} being replaced by $\mu_y{}^2 \delta_x{}^2 u/(\Delta x)^2$ while the cross derivative is treated in the same way as in (10).

Straight forward algebraic manipulation confirms that

$$\sum_{i=1,j=1}^{I-1,J-1} \underline{u}_{ij} \cdot \underline{\nabla}_\Delta a_{ij} + \sum_{i=o,j=o}^{I-1,J-1} a_{i+\frac{1}{2}\,j+\frac{1}{2}} \, \underline{\nabla}_\Delta \cdot \underline{u}_{i+\frac{1}{2}\,j+\frac{1}{2}} = \text{boundary terms,} \tag{21}$$

which may be written concisely as

$$\langle \underline{u}, \underline{\nabla}_\Delta a \rangle_{xv} + \langle \underline{\nabla}_\Delta \cdot \underline{u}, a \rangle_{\cdot s} = o \tag{22}$$

under appropriate boundary conditions : and similarly we have

$$\langle \underline{u}, \underline{\nabla}_\Delta \times \underline{v} \rangle_{xv} - \langle \underline{\nabla}_\Delta \times \underline{u}, \underline{v} \rangle_{\cdot v} = o. \tag{23}$$

In these identities $\underline{u}$ is defined on the **x** points and a and $\underline{v}$ on the · points.

Applying this scheme to (19) is obvious. For the full equations of (A1) however application is a little more complicated: the other second order term, $\underline{J}' \times \underline{B}_o$, is straightforward but as is shown in the Appendix the two first order terms $\underline{\nabla}(\underline{\xi} \cdot \underline{\nabla} p_o)$ and $\underline{J}_o \times \underline{B}'$ are only together self-adjoint by virtue of the equilibrium equation (1). This property must also be exactly preserved in the difference scheme : recognition of this fact then makes implementation possible.

The stability of the difference schemes is treated most naturally by the energy method. Having established conservation of an appropriately defined energy given by (16), it is merely necessary to obtain the conditions under which it is postive definite. For example, the argument for the reduced equation (19) is as follows:

$$\langle \underline{\nabla}_\Delta \cdot \underline{\xi}^n , \Gamma \underline{\nabla}_\Delta \cdot \underline{\xi}^{n+1} \rangle_{\cdot s} = \| \Gamma^{\frac{1}{2}} \underline{\nabla}_\Delta \cdot \underline{\xi}^n \|^2_{\cdot s} + \langle \underline{\nabla}_\Delta \cdot (\underline{\xi}^{n+1} - \underline{\xi}^n), \Gamma \, \underline{\nabla}_\Delta \cdot \underline{\xi}^n \rangle_{\cdot s} .$$

The first term on the right and the kinetic energy are positive definite and must dominate the second term. This is bounded in magnitude by

$$\frac{\Delta t}{\Delta x}\left[\frac{\Gamma}{\rho_o}\right]^{\frac{1}{2}}_{max} \left|\left\langle \rho^{\frac{1}{2}}(\Delta x \quad \underline{\nabla}_\Delta)\cdot \frac{\underline{\xi}^{n+1} - \underline{\xi}^n}{\Delta t} \quad , \Gamma^{\frac{1}{2}} \underline{\nabla}_\Delta\cdot\underline{\xi}^n\right\rangle_{\cdot s}\right|$$

$$\leqslant \quad \frac{\Delta t}{\Delta x}\left[\frac{\Gamma}{\rho_o}\right]^{\frac{1}{2}}_{max} \left[\left\|\frac{\underline{\xi}^{n+1} - \underline{\xi}^n}{\Delta t}\right\|^2_{xv} + \left\|\Gamma^{\frac{1}{2}} \underline{\nabla}_\Delta\cdot\underline{\xi}^n\right\|^2_{\cdot s}\right]$$

so that stability follows if $(\frac{\Delta t}{\Delta x})^2 \; \Gamma/\rho_o < 1$.

Boundary Conditions

The boundary terms occurring in (21) consist of all those terms in the second sum which involve $\underline{u}_{i,j}$ with $i = o$ or I or $j = o$ or J. Because the $\times$ mesh and $\cdot$ mesh imply the approximation of the volume integral by sums over two different cell systems, there are several possible interpretations of the boundary terms : they lead to differing definitions of energy sums of (16) and differing implementations of the boundary condition $\xi_n = 0$. A fully consistent, zero order accuracy choice for the reduced equation (19) is as follows.

For a rectangular region with boundaries at $i = o, I$ and $j = o, J$ the $\cdot$ sum for the potential energy $\Gamma|\underline{\nabla}\cdot\underline{\xi}|^2$ of (19) is taken over all interior $\cdot$ points. The x sum for the kinetic energy is then taken over all interior $\times$ points plus one half of the sum for the boundary points. On the boundary ξ_n is put to zero ($\underline{\xi} = o$ at the corners) and the tangential component stepped forward in a special way: $\underline{\nabla}_\Delta\cdot\underline{\xi}$ is defined at all $\cdot$ points and for an interior $\times$ point four $\cdot$ neighbours are involved in forming $\underline{\nabla}_\Delta(\underline{\nabla}_\Delta\cdot\,\underline{\xi})$; but at the boundary only the tangential derivative is required and this is obtained from the immediately interior neighbours as indicated in Figure II. The lack of averaging here is compensated by the half contribution to be taken in the boundary sums of equation (15) and results in exact energy conservation.

A general curved boundary is approximated by a rectilinear boundary through $\times$ points. "Straight line" points like a, c and f in Figure III are treated exactly as before but corner points must be treated more carefully.

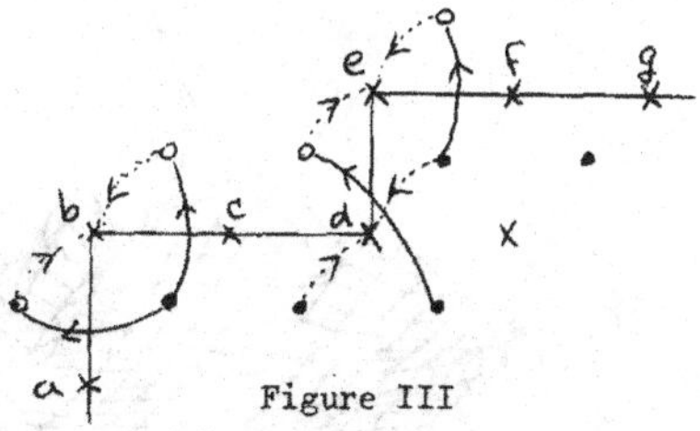

Figure III

Again the normal component of $\underline{\xi}$, that is $\xi_x \pm \xi_y$, is put to zero and the tangential component stepped forward by a single difference between a pair of $\cdot$ points. For an "interior corner",like d, the most appropriate diagonal interior points are available as shown: but for an "exterior corner", like e, they must be extrapolated - along the y - mesh if the corresponding neighbour, f, is a "straight x - line" point or diagonally if the neighbour, d, is a corner point.

Correspondingly, interior corners are fully counted in the kinetic energy sum and exterior corners ignored: calculation of $\underline{\xi}$ at the latter is merely necessary for calculating $\underline{\nabla}_{\Delta}\cdot\underline{\xi}$ at the next time step. These definitions again lead to exact energy conservation.

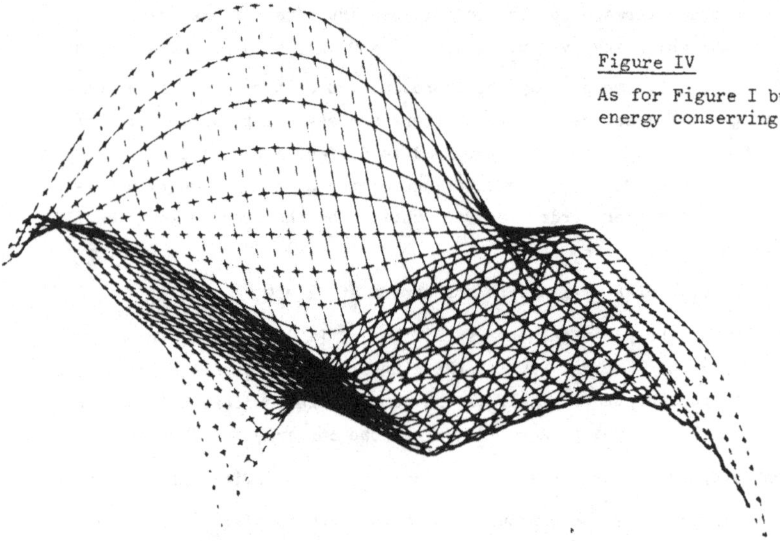

Figure IV

As for Figure I but with energy conserving scheme.

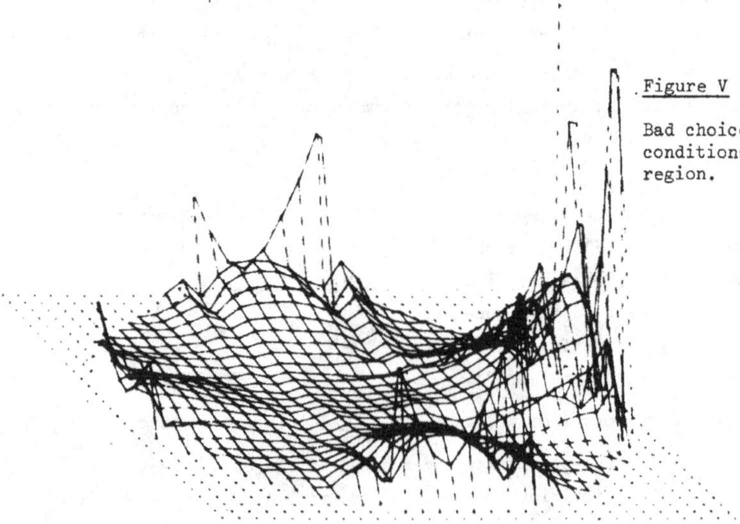

Figure V

Bad choice of boundary conditions on a circular region.

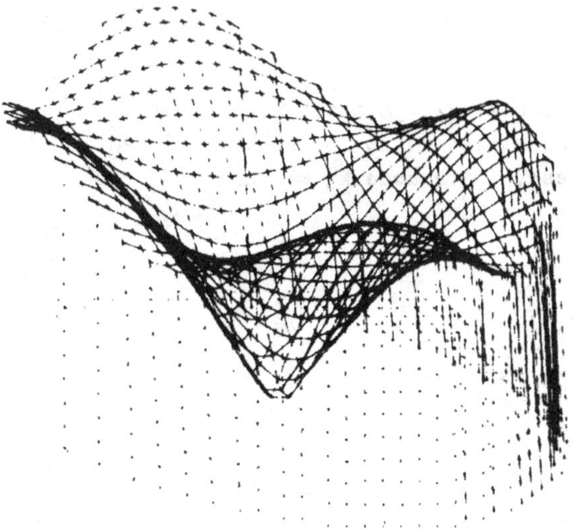

Figure VI

Good boundary conditions on a circular region.

Typical Results

Most of the applications of interest are in cylindrical co-ordinates where the region studied is a cross-section of an axially symmetric torus. Practical calculations are still at an early stage and in any case are best presented in a short movie. However, Figures IV, V, VI and VIII give some indication of the effectiveness of the method. Figure IV shows the same calculation as Figure I but using the energy concerning difference scheme and boundary conditions (on a rectangle). Figure V demonstrates the dangers of badly chosen but not unreasonable boundary conditions on a circle, while Figure VI shows the value of a careful choice. Finally, Figure VII shows how with only a 20 x 20 mesh the calculated growth rate varies with the major radius of the torus, until in the infinite limit it can be compared with an analytic result.

The results are encouraging and show clearly the value of a careful design of difference scheme, especially in the spatial variables. What is less certain is whether straight forward solution of the initial-boundary value problem is the best approach. It may well transpire that these spatial difference schemes should be used in an attack on the eigenvalue problem for the fastest growing mode.

Figure VII

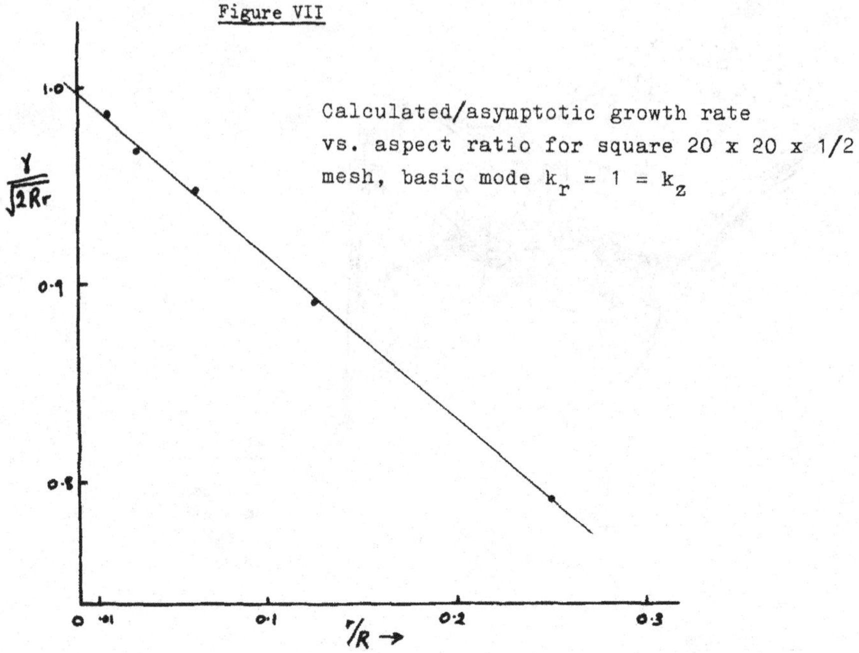
1.0
0.9
0.8
$\frac{\gamma}{\sqrt{2Rr}}$
0
.01
0.1
0.2
0.3
r/R →
Calculated/asymptotic growth rate
vs. aspect ratio for square 20 x 20 x 1/2
mesh, basic mode $k_r = 1 = k_z$

References

1. Courant, R. Friedrichs, K.O. & Lewy, H. Über die partiellen Differenzengleichungen der mathematischen Physik. Math. Ann 100 p.32, 1928.
2. Glastone, S. & Lovberg, R.H. Controlled Thermonuclear Reactions (Chap. XIII Plasma Stability Theory). Van Nostrand, 1960.
3. McKee, S. Alternating direction methods for hyperbolic equations with mixed derivatives. J.Inst. Maths. Applics. 9 p.350, 1972.
4. Morton, K.W. The design of difference schemes for evolutionary problems. SIAM-AMS Proc., 2 p.1, 1970.

Appendix

The M.H.D. Perturbation Equations

Static equilibrium is described by a density ρ_o, pressure p_o, magnetic field $\underline{B}_o$ and current $\underline{J}_o = \frac{c}{4\pi}\, \underline{\nabla} \times \underline{B}_o$. Perturbation quantities ρ', p', $\underline{B}'$ and $\underline{J}'$ can all be expressed in terms of a displacement $\underline{\xi}$ with the flow velocity $\underline{u}' = \partial\underline{\xi}/\partial t$. Then the equation of motion becomes

$$\rho_o \frac{\partial^2 \underline{\xi}}{\partial t^2} = - \underline{\nabla}\, p' + \frac{1}{c}\left[\underline{J}' \times \underline{B}_o + \underline{J}_o \times \underline{B}'\right] \qquad \text{(A1)}$$

where

$$p' = - \gamma\, p_o\, \underline{\nabla}\cdot\underline{\xi} - (\underline{\xi}\cdot\underline{\nabla})p_o$$

$$\underline{B}' = \underline{\nabla} \times (\underline{\xi} \times \underline{B}_o)$$

and

$$\underline{J}' = ({}^{c}/4\pi)\; \underline{\nabla} \times \underline{B}'$$

Energy conservation is demonstrated by taking the scalar product of the equation with $\partial\underline{\xi}/\partial t$ and integrating over the whole volume. The left-hand side gives

$$\text{change in kinetic energy,} \quad \frac{\partial}{\partial t}\, \tfrac{1}{2} \int \rho_o \left|\frac{\partial \underline{\xi}}{\partial t}\right|^2 dV. \qquad \text{(A2)}$$

The first term on the right gives, as its first part,

$$\int \frac{\partial \underline{\xi}}{\partial t} \cdot \underline{\nabla}(\gamma\, p_o \underline{\Delta}\cdot\underline{\xi})\, dV = - \frac{\partial}{\partial t}\, \tfrac{1}{2} \int \gamma\, p_o (\underline{\nabla}\cdot\underline{\xi})^2\, dV + \int \gamma\, p_o \frac{\partial \xi_n}{\partial t}\, \underline{\nabla}\cdot\underline{\xi}\, dS. \qquad \text{(A3)}$$

Similarly, the $\underline{J}' \times \underline{B}_o$ term gives

$$\frac{1}{4\pi} \int \frac{\partial \underline{\xi}}{\partial t} . (\underline{\nabla} \times \underline{B}') \times \underline{B}_o\, dV = - \frac{1}{4\pi} \int \left(\frac{\partial \underline{\xi}}{\partial t} \times \underline{B}_o\right) . (\underline{\nabla} \times \underline{B}')\, dV$$

$$= - \frac{\partial}{\partial t} \int \frac{|\underline{B}'|^2}{8\pi}\, dV - \int \left[\left(\frac{\partial \underline{\xi}}{\partial t} \times \underline{B}_o\right) \times \underline{B}'\right]_n dS. \qquad \text{(A4)}$$

However, the two first order terms $\nabla(\underline{\xi}\cdot\underline{\nabla}\, p_o) + \underline{J}_o \times \underline{B}'/c$ must be treated together and the equilibrium equation made use of to show that, when $\xi_n = o$ on S, the right hand side of (A1) gives minus the expression

$$\frac{\partial}{\partial t}\, \tfrac{1}{2} \iint \left[\gamma\, p_o (\underline{\nabla}\cdot\underline{\xi})^2 + (\underline{\nabla}\cdot\underline{\xi})\underline{\xi}\cdot\underline{\nabla}\, p_o + \frac{|\underline{B}'|^2}{4\pi} - \frac{\underline{J}_o \cdot \underline{B}' \times \underline{\xi}}{c}\right] dV, \qquad \text{(A5)}$$

i.e., the change in potential energy. A convenient reference for plasma stability theory as used here is Glastone and Lovberg(2).

VARIABLE ORDER VARIABLE STEP FINITE DIFFERENCE METHODS FOR NONLINEAR BOUNDARY VALUE PROBLEMS

VICTOR PEREYRA

1. INTRODUCTION

We intend to present in this paper a concise survey of some high order finite difference methods based on deferred corrections. The theoretical foundation, many of the tools, and some of the applications, have been developed in the last 8 years, and because of some recent results it seems a fitting time to recapitulate.

Basic notation and results are presented in Section 2 in a general though somewhat informal way.

The problem to which this work is addressed is stated in Section 3. In short it says:"Given a boundary value problem, a basic mesh region Ω_h, a discretization, and a tolerance, find an approximate solution defined (at least) in Ω_h and within the tolerance". A general scheme for solving this problem is also described in Section 3. Section 4 deals with applications and numerical results concerning nonlinear two-point boundary value problems, multipoint boundary value problems for first order nonlinear piecewise smooth systems, and finally some two-dimensional elliptic boundary value problems. In this last application the novelty lies in the coupling of fast direct methods with deferred corrections.

2. BASIC RESULTS AND NOTATION

In this Section we endeavor to present basic facts, that are already well known and will be needed subsequently, in a sufficiently general but simple notation which will serve more as a frame of reference than as an ultraformal hindrance. Thus we will not strive either to introduce the most general spaces and conditions possible or to be absolutely precise about all the details. There are two reasons for this choice of style: one is that the precise functional analysis formulations have been given earlier (Pereyra (1967b,c)), and the other is that in this paper we are mostly interested in calling the reader's attention to implementation and numerical performance.

We want to consider various boundary value problems for nonlinear differential equations, which we shall represent as

(2.1) $$F(y) = 0$$

with the domain and range of the operator F contained in appropriate function spaces. This is the continuous problem which, from now on, is assumed to have an unique solution y^*.

Finite differences will be used to provide discrete approximations Y defined only on a finite mesh region. A positive parameter h will be associated with the mesh, measuring in some way its maximum diameter. As usual, h is meant to tend to zero, and through the whole discussion we will consider $0<h<h_0$, for some fixed h_0.

The discrete problem will be denoted by

(2.2) $$F_h(Y) = 0.$$

In order to connect the continuous and discrete spaces of unknowns we introduce, for each h, a discretization mapping ϕ_h. With this mapping we can define the local truncation error $\tau_h(y^*)$:

(2.3) $$\tau_h(y^*) \equiv F_h(\phi_h y^*).$$

We assume that a unique solution Y(h) to (2.2) exists for each h.

We shall say that Y(h) converges discretely to y^* iff

(2.4) $$\lim_{h\to 0} \| Y(h)-\phi_h y^* \| = 0.$$

The discretization is consistent of order $p>0$ iff $\tau_h(y^*) = O(h^p)$. Stability is a notion that is inseparable from the two just given. It is well known that in the linear case stability of a discretization is simply implied by the uniform boundedness (with respect to h) of the inverse operator. In the nonlinear case a convenient way of stating the stability condition is the following: F_h is stable iff for any pair of discrete functions U,V, there exists a constant $c>0$, independent of h, such that

(2.5) $$\| U-V \| \leq c \| F_h(U)-F_h(V) \| .$$

A fundamental result (see for instance Pereyra (1967b) or (1973)) is the following:

Theorem 2.1 Let F_h be stable on the spheres $B_h \equiv B(\phi_h y^*, \rho)$ and consistent of order p with F. Then there exists an $h_0>0$ such that for each $0<h<h_0$ the discrete problem has an unique solution Y(h) which is convergent of order p to y^*, i.e. $\| Y(h)-\phi_h y^* \| = O(h^p)$.

The vector $e(h) = Y(h)-\phi_h y^*$ is called the global truncation error.

If a more detailed expression for the local truncation error is available in the form of an asymptotic expansion in powers of h then it is possible, under reasonable conditions, to obtain similar asymptotic expansions for the global error. In fact, let us assume that τ_h

can be expressed as

$$(2.6) \qquad \tau_h = \sum_{j=p}^{J} F_j(y^*)h^j + O(h^{J+1}),$$

where the F_j are known operators (usually combinations of high order derivatives of y^*). Then, essentially under the assumptions of Theorem 2.1 and suitable smoothness requirements, there exists an expansion for $e(h)$:

$$(2.7) \qquad e(h) = \phi_h \sum_{j=p}^{J} e_j h^j + O(h^{J+1}),$$

where the functions e_j are solutions of appropriate linear boundary value problems (see Stetter (1965), Pereyra (1967b,c), Lentini (1973) for proofs of this result in a general setting). If expansion (2.6) is lacunary in some regular way, for instance if only the terms F_{kp} are different from zero, then it is possible to show that expansion (2.7) will have a similar structure. Expansion (2.7) is basic for the method of Successive Extrapolations (Richardson extrapolation to h=0, Romberg integration), and in the survey papers of Joyce (1971) and Widlund (1971) a wealth of information and references about this important technique can be found. Given an initial mesh size h_0 the method of Successive Extrapolations (SE) requires the computation of the discrete solutions $Y(h_i)$, where h_i, $i=1,\ldots,I$, is a decreasing sequence such that the corresponding meshes M_i are refinements of the initial mesh M_0 associated with h_0. Appropriate linear combinations of the values of the mesh functions at the common grid points will produce the successive elimination of the leading error terms in (2.7). Accurate discrete solutions are then obtained on the coarsest mesh M_0 at the cost of solving equation (2.2) on much finer meshes. For one equation in one dimension this added cost can probably be offset by the simplicity of the algorithm. For systems of equations or in higher dimensions the storage and time requirements grow considerably and therefore more efficient techniques become necessary.

2.1. Deferred corrections.

As early as 1947 Leslie Fox advocated a technique for increasing the accuracy of discrete solutions that he called "difference corrections". Through the years he and his collaborators have applied this technique to a variety of problems in differential and integral equations as is attested in the book edited by Fox (1962). See also Pereyra (1967c) for an extensive set of references and historical account.

In Fox (1962) we find the term "deferred corrections" used interchangeably with that of difference corrections. We have preferred to adopt the former in our work which is, at least in appearance, differ-

ence free.
Deferred corrections are based on the observation that heightened precision can be obtained from a lower order solution by approximating terms in the local truncation error and re-solving. The important features are that the <u>same</u> grid size h is used throughout, and the <u>same</u> low order, simple operator must be inverted at each step. Also, as is the case of Successive Extrapolations, an algorithm for estimating asymptotically the global truncation error is implicit in the method.

Let T_ℓ be the segment of the local truncation error containing the first ℓ terms in the sum (2.6). Let S_ℓ be an approximation to T_ℓ (on the grid points) satisfying

(2.8) $$T_\ell \equiv \phi_h \sum_{j=p}^{\ell.p} F_j(y^*)h^j = S_\ell(\phi_h y^*)+O(h^{(\ell+1).p}).$$

Then the <u>deferred correction</u> procedure is described by:

0) Solve $F_h(Y) = 0$. Call $Y^{(0)}$ its solution.

k) Solve $F_h(Y) = S_k(Y^{(k-1)})$. Call its solution $Y^{(k)}$. Repeat for k=1,...

In Pereyra (1967b,c) it is proven in a fairly general setting that the mesh functions $Y^{(k)}$ determined by this procedure satisfy

(2.9) $$Y^{(k)}-\phi_h y^* = O(h^{(k+1).p}).$$

As a matter of fact, the equations need not be solved exactly (an impossible feat in practice), and that is also contemplated in the results mentioned above.

Naturally, the practical application of this procedure depends upon our ability to obtain the approximations S_k to the local truncation error. In all of our implementations we have made the natural choice, i.e. since the terms of the expansion for the local error are high order derivatives of the solution y*, we consider numerical differentiation formulas in terms of function values at the mesh points. As we shall see, very often the derivative order and the power in h coincide for each term in the sum to be approximated and therefore one of the dangers of numerical differentiation is not present. In true fact we are approximating differences of y* (remember Fox !), but our approach gives us a great deal of flexibility in automating all the procedures in a very efficient and accurate way. In Björck and Pereyra (1970), Galimberti and Pereyra (1970, 1971) and in Pereyra and Scherer (1973) a variety of algorithms and software are developed to cope, in particular, with numerical differentiation in one and several variables, both with Lagrangian and Hermite formulas. Incidentally, some of those algorithms could be used with profit in other tasks relevant to this Conference, notably the computation of finite elements and of weights for numerical quadratures.

We would like to stress that formula (2.9) shows a remarkable property of deferred corrections, i.e. for a method of order p, p orders can be gained per correction. This fact was first observed by Daniel (see Daniel, Pereyra and Schumaker (1968)), and it can be seen at work in Pereyra (1968, 1973) , and in Section 4.1 of this paper, where a method of order four is used for two-point boundary value problems. A generalization to mildly nonlinear equations on rectangular regions using as a basic method a nine point formula and a newly developed fast direct solver is being completed (Concus, Golub and Pereyra (1973)).

For nonlinear problems, if it is desired to perform only one correction then a very economical procedure is available, specially if the nonlinear discrete equations are being solved by Newton's method. In fact, in that case, a program module should be available for computing the Jacobian matrix $F_h'(Y)$ and for solving linear systems of the form $F_h'\Delta = \beta$. It turns out that the following procedure will double the asymptotic order of our algorithm by a price equivalent to one Newton step and the computation of S_1:

0) Solve $\quad F_h(Y) = 0 \quad$ for $\quad Y^{(0)}$.

1) Solve the linear problem $F_h'(Y^{(0)})\Delta = S_1(Y^{(0)})$.

2) Correct: $Y^{(1)} = Y^{(0)} - \Delta$.

Then $\quad Y^{(1)} - \phi_h y^* = O(h^{2p})$.

2.2. Asymptotic error estimates via deferred corrections.

Whenever a method to improve the accuracy exists, not far behind there must be one to estimate the error of the unimproved solution. In our case it is possible to prove (Pereyra (1970, 1973)) that the solution Δ_k to the linear problem

$$F_h'(Y^{(k)})\Delta = S_{k+1}(Y^{(k)}) - S_k(Y^{(k-1)}) \tag{2.10}$$

satisfies

$$\phi_h y^* - Y^{(k)} = \Delta_k + O(h^{p(k+1)}). \tag{2.11}$$

In other words, Δ_k is an asymptotic estimator for the error in $Y^{(k)}$ (or it can be used as a correction if that is the last step to be performed as we did in the last Section).

This error estimator can be used in a variety of ways besides the obvious one of reporting on line the accuracy of the computed solutions. In the applications to be described below we compare $\|\Delta_k\|$ with

$||\Delta_{k-1}||$ in order to take decisions about continuing corrections, decreasing the step, or interrupting the computation. It is clear that, since the vector Δ_k gives precise information about the error at each mesh point, that local step adjustments could be based on it.
The knowing reader will be aware by now that many of these ideas have been implicitly or explicitly advocated by Dahlquist and Henrici (1962), though with the exception of the initial value problem for ordinary differential equations we know of no systematic use of them in practical applications. The aim of this work is to call attention upon the possibility of constructing software for boundary value problems of similar characteristics to that available for initial value problems (cf. Gear (1971), Krogh (1969a, 1969b), and Bulirsch and Stöer (1966) as implemented by N. Clark (see Crane and Fox (1969))), and we shall proceed to expound in the next Sections some of the results of our efforts.

3. A GENERAL SCHEME

Based on the notation developed in Section 2 we present now a skeleton of the algorithm used in most of the applications to be described in Section 4. The problem to which the algorithm is addressed is the following:
"Given problem (2.1), a mesh region Ω_h (described shortly by the parameter h), a tolerance TOL>0, and discretization (2.2), find an approximate solution Y defined (at least) on Ω_h satisfying $||Y-\phi_h y^*|| <$TOL".
The basic mesh region Ω_h is that in which the user wants the approximate solution to be defined (minimal description).

Starting with Ω_h and the basic discrete method (2.2) an approximate solution is computed and an asymptotic error estimate is produced. If necessary, the net would have been refined in order to make that first step possible. Whenever the tolerance TOL is met, according to the error estimate, the process is interrupted. If the required accuracy has not been achieved then (2.2) is re-solved with the already computed correction vector $S_1(Y^{(0)})$ added in the right hand side. $S_2(Y^{(1)})$ is computed and used to produce a new error estimate, which is then compared with the former step error estimate. In order to continue the corrections on this mesh it is required that a significant gain in accuracy had been obtained, otherwise the mesh is refined. This last step acts exactly as the traditional sentinel on "the growth of high order differences", except that it takes into account other more complicated factors and that, of course, no differences must ever be inspected.

Whenever the mesh is refined the process restarts, with the difference that a fairly precise initial approximation is obtained by interpolation of the available approximate solution on the coarser mesh. That this procedure is feasible and effective in a variety of non-trivial applications is borne out by the numerical results presented in the next Section and in the references.

4. APPLICATIONS

4.1. Special two-point boundary value problem.

The simple problem

$$y''-f(x,y) = 0$$
$$y(a) = \alpha,\ y(b) = \beta \tag{4.1}$$
$$f_y > -(\Pi/(b-a))^2$$

is one of the battle horses of many papers when testing hour arrives. Though the results of Section 4.2 will supersede completely those presented here, for historical and comparative reasons we would like to mention some of the results obtained with an implementation of the fourth order method

$$h^{-2}(-Y_{i-1}+2Y_i-Y_{i+1})+\frac{1}{12}(f_{i-1}+10f_i+f_{i+1}) = 0\ , \quad i=1,\dots,n-1$$
$$Y_0 = \alpha,\quad Y_n = \beta, \tag{4.2}$$

on the uniform mesh $x_i = a+ih$, $i=0,\dots,n$, $h = (b-a)/n$. For smooth f the asymptotic expansion for the local truncation error is

$$\tau_h(y^*) = \phi_h \sum_{j=2}^{J} a_j f^{(2j)}(x,y^*(x))\frac{h^{2j}}{(2j)!} + O(h^{2J+2}) \tag{4.3}$$

with $\quad a_j = \frac{1}{(j+1)(2j+1)} - \frac{1}{6}$.

At essentially the same cost as the simplest $O(h^2)$ method an $O(h^8)$ method can be obtained by approximating the first two terms in the sum (4.3) and performing one linearized deferred correction as indicated at the end of Section 2. Details on how the correction terms are actually computed are beyond the scope of this paper but have been given earlier in Pereyra (1965, 1967a, 1968, 1973).

An adaptive method (M2) based on the discretization (4.2) and following the scheme described in Section 3 has also been implemented.

Method M3 is an implementation of successive extrapolations. More details and Fortran programs for these methods are available in Pereyra (1973).

Our numerical experience indicates that these methods should be preferred (when they apply) to the general method of Section 4.2, since they are usually more efficient and economical.

Example 1: $-y''+e^y = 0$,

$y(0) = y(1) = 0$.

Exact solution: $y^*(x) = -\ln 2+2 \ln (c \sec(\frac{c}{2}(x - \frac{1}{2})))$,
where the constant c satisfies: $c \sec \frac{c}{4} = \sqrt{2}$. The constant c is to 16 significant digits c = 1.336055694906108...

Whenever possible, we shall report the most accurate results obtained, which will be limited by the word length of the computer used, i.e. ∿16 decimal digits on IBM System 360 "double precision". Paraphrasing Keller (1972), this by no means signifies that our algorithms are only useful for these highly accurate solutions, since more modest accuracy can be obtained very economically on fairly coarse meshes.

The present problem has been chosen because of its frequent appearance in the literature of high order methods (cf. Perrin, Price, and Varga (1969), Ciarlet, Schultz, and Varga (1967), Herbold, Schultz, and Varga (1969), Jerome, and Varga (1969), Keller (1972), and Pereyra (1973)).

TABLE 1

Method	Initial mesh	Final mesh	Exact max. abs. error	Theoretical final order of the method
M1	32	32	4.08,-15	8
M2	8	16	5.2 ,-15	12
M3	8	64	2.5 ,-16	10

We point out that the estimated max. abs. error for M2 was equal to 5.5,-15, predicting very accurately the true error.

4.2. Multipoint boundary value problems for nonlinear first order systems.

Following Keller (1969, 1972) we consider multipoint boundary value problems of the form

$$(4.4) \qquad y'(t)-f(t,y(t)) = 0 , \quad a<t<b, \qquad g(y(\tau_1),\dots,y(\tau_N)) = 0 ,$$

where y, f, and g are n-dimensional vector functions. Under suitable conditions Keller has proven that if $y^*(t)$ is an isolated solution of (4.4) then the discrete scheme known as the Box or centered-Euler scheme will provide an $O(h^2)$ approximation which can be computed by Newton's method. The error will have an asymptotic expansion in even powers of h, provided the data is piecewise smooth and the jump discontinuities occur at the boundary points. These results apply as well to the trapezoidal rule

$$\frac{1}{h_j}(Y_j - Y_{j-1}) - \frac{1}{2}(f(t_{j-1}, Y_{j-1}) + f(t_j, Y_j)) = 0 \tag{4.5}$$

which is the one we have chosen in our experimentation. The work I am reporting in this Section is joint work with M. Lentini (cf. Lentini and Pereyra (1973), and Lentini (1973) where more details can be found). The nets considered by Keller are completely non-uniform, and eventually we could implement a variable order, variable step, deferred correction procedure of that generality. Since our work here constitutes one of the first steps in this direction (cf. Keller (1969), p. 22) and since further results require a major programming effort, we have limited somewhat the generality to the following simpler problem (cf. Keller (1968)):

$$\begin{aligned} & y'(t) - f(t, y(t)) = 0 \;, \\ & Ay(a) + By(b) = \alpha \;, \end{aligned} \tag{4.6}$$

for which we use only uniform meshes.

However, we shall also present some preliminary results for a case in which there is an internal point of discontinuity, showing clearly that deferred corrections can be applied in such situations. This is a major departure from earlier results and we expect to pursue it further. It is clear now that going from an implementation for (4.6) to one for (4.4) is not too difficult. We doubt that completely general meshes are called for, and probably we will only consider piecewise uniform ones, i.e. independent uniform subdivisions between consecutive boundary (or discontinuity) points. In such a case our program is practically the same we have at present, with a minor modification in the linear equation solver and an extra loop in the correction generator. Provided the corrections are made using information only on these subintervals, without straddling boundary points, there are no new difficulties in the theoretical part either.

The asymptotic expansion for the local truncation error corresponding to the discretization (4.5) is

$$\tau_h = - \sum_{j=1}^{J} \frac{j}{2^{2j-1}(2j+1)} f^{(2j)}(x, y^*(x)) \frac{h^{2j}}{(2j)!} + O(h^{2J+2}) \;. \tag{4.6'}$$

With a minor modification due to the fact that the values of Y at the end points are also unknown, the deferred correction generator of Pereyra (1973) can, and has been used, in our implementation.

It would be very advisable to develop a different program for second order systems when no first derivatives are involved, generalizing the method described in Section 4.1.

We remark that high order equations can be reduced to first order systems in the usual way, but that in sharp contrast with other high order methods, both the function and all its derivatives up to one unit less than the order of the equation are obtained with the same accuracy.

Example 2 (Ciarlet, Schultz, and Varga (1967)).

$$y^{(IV)}(x) = (x^4+14x^3+49x^2+32x-12)e^x ,$$

$$y(0) = y'(0) = y(1) = y'(1) = 0 .$$

Exact solution: $y^*(x) = x^2(1-x)^2e^x$.

Starting with a uniform grid containing 9 points (including the boundary points) automatic subdivision and corrections furnished the following results:

TABLE 2

	Exact max. error in y(x)	Exact max. error in y'(x)
Ciarlet, Schultz, Varga (1967) (best results)	7.614,-5	2.130,-3
This method 9 points 2 correct.	1.975,-5	2.893,-5
This method 33 points 6 correct.	2.162,-15	1.443,-15

Again the errors were estimated very precisely.

Example 3. A problem with one jump discontinuity (Perrin, Price, and Varga (1968)):

$$y^{(IV)}(x) = \begin{cases} 24.0 , & 0\le x\le 1/2 , \\ 48.0 , & 1/2<x\le 1 , \end{cases}$$

with the same boundary conditions of Example 2.

Exact solution:

$$y^*(x) = \begin{cases} x^4 - \frac{19}{8}x^3 + \frac{21}{16}x^2 , & 0\le x\le 1/2 , \\ 2(x-1)^4 + \frac{29}{8}(x-1)^3 + \frac{27}{16}(x-1)^2 , & 1/2<x\le 1 . \end{cases}$$

In Perrin, Price, and Varga (1968) a variational method using cubic Hermite test functions is compared with a finite difference method that consists of a five points, $O(h^2)$ discretization of $y^{(IV)}(x)$. The variational method shows beautifully the expected fourth order approximation, while the finite difference method shows a very erratic behavior. The best results for the variational method are obtained with a Hermite subspace of dimension 18 and the max. abs. error on [0,1] is 1.25×10^{-5}, while for the finite difference method the best results on a mesh with 80 points give a max. abs. error on the grid points of 8.53×10^{-3}. These results seem to discredit all finite difference methods.

However, it is clear from our brief discussion earlier that (a) the finite difference approximation straddles the discontinuity for grid points in a neighborhood of $x = 1/2$, while the two point Hermite approximation does not, (b) in the best of cases, the finite difference approximation will only have order $O(h^2)$.

What happens if one uses an appropriate (though not special!) finite difference method?

In Table 3 we present the results obtained with our adaptive method. Again we list max. abs. errors on the grid points and also for each correction, the computed order (O.) of the method obtained comparing the errors for two successive meshes. N is the total number of mesh points, including the end points, and k is the correction number.

TABLE 3

N\k	0	O.	1	O.	2	O.	3	O.
9	6.05,-3	-	4.43,-6	-	-	-	-	-
17	1.53,-3	2.0	2.75,-7	4.0	1.08,-9	-	4.22,-12	-
33	3.82,-4	2.0	1.72,-8	4.0	1.68,-11	6.0	1.65,-14	8.0
65	9.56,-5	2.0	1.07,-9	4.0	2.62,-13	6.0	6.94,-17	7.9

Thus we see that finite differences, properly used, can handle this type of problem and can produce truly highly accurate solutions. Also it is important to remark that again the error in the approximations to $y^{(i)}$, $i = 0,1,2,3$ was predicted very accurately and that all these derivatives were approximated with the same relative error. Observe that the method without corrections (column 0) shows a perfect non-erratic behavior, as predicted by the theory. (So do the corrected ones by the way.) These results were obtained in 78.17 sec. of computing time on an IBM 360/50 computer, using 138 K bytes of main memory. By eliminating the printing this time can be considerably reduced.

4.3. Poisson, Helmholtz and mildly non-linear elliptic equations. Rectangular regions.

The linear equations considered are

(4.7a) $-\Delta u+f(x,y) = 0$,

(4.7b) $-\Delta u+\lambda u+f(x,y) = 0$,

with Dirichlet boundary conditions on a rectangle of the plane (x,y).

The Laplacian operator is discretized on an uniform mesh by the usual 5-point formula (Δ_h).

Combining the fast linear equation solver of Buneman (1969) with deferred corrections provides a very effective, highly accurate procedure for these multidimensional problems.

The asymptotic expansion corresponding to Δ_h is simply (cf. Pereyra (1967a))

(4.8) $$\tau_h = \sum_{j=1}^{J} \frac{2}{(2j+2)!} \left(\frac{\partial^{2j+2}u}{\partial x^{2j+2}} + \frac{\partial^{2j+2}u}{\partial y^{2j+2}}\right)h^{2j}+O(h^{2J+2}) \ .$$

Again the deferred correction generator of Pereyra (1973) can be used for obtaining the corrections in each coordinate direction (cf. Pereyra (1970) for more details and generalizations).

Example 4.

$-\Delta u-50 \sin(5(x+y)) = 0$, on the unit square.

Exact solution (and boundary function)

$u^* = \sin(5(x+y))$

We give in Table 4 the exact and estimated max. relative errors on a 32x32 grid for several corrections.

TABLE 4

Correction	estimated max. rel. error	exact max. rel. error
0	0.698,-1	0.652,-1
1	0.936,-4	0.809,-4
2	0.119,-4	0.127,-4
3	0.922,-6	0.844,-6
4	0.379,-7	0.775,-7

These results were obtained on the IBM 360/67 computer at the Stanford University Computing Center. The execution time was 49.94 seconds, using a WATFIV program. Observe that the factor 5 in the argument of the sin function gives a solution with large derivatives. For instance if we consider the problem $-\Delta u-2 \sin(x+y) = 0$, whose solution is $u^*(x,y) = \sin(x+y)$, then in four corrections a precision of 0.46×10^{-11} is achieved with the same computing time. The five point formula alone (i.e., with no corrections) on a 64x64 grid gives in this case a precision of 0.29×10^{-5} in 17.12 seconds of computing time. Due to memory restrictions the largest mesh that could be processed with this program is 512x512. Extrapolating from the results for coarser meshes the $O(h^2)$ method would give an accuracy of $.46\times10^{-9}$ in about 1649 seconds of computing time. Results for the Helmholtz equation (4.7b) are similar. Details of this work will be published elsewhere. Coupling the techniques of Concus and Golub (1972) with deferred corrections we can extend these results to mildly nonlinear elliptic equations. The nonlinear equations are solved via a relaxed Picard iteration and a fast Helmholtz solver. Preliminary experimentation seems to indicate that in this case it is better to use a fix length correction formula from the beginning in order to diminish the number of nested interations. It would be interesting to use a Newton type iteration, but in that case the use of fast solvers seems to be precluded. Instead of using the five point formula as a basic method one can use an $O(h^4)$ nine point formula (cf. Kantorovich and Krylov (1958)), for which a fast solver has just been developed by G.H. Golub. In this joint work with P. Concus and G.H. Golub we plan to extend to two dimensions the method of Section 4.1.

4.4. <u>Poisson's equation on curved regions</u> (Pereyra and Widlund (1973)).

For general regions the theoretical and practical difficulties of high order finite difference methods are considerably increased. To start with, the existence of asymptotic expansions depends very much upon the treatment of the boundary (cf. Pereyra (1967a)). A few years ago H.O. Kreiss proposed a method which, by using high order Lagrange interpolation at the boundary would guarantee the existence of expansions in even powers of h up to order h^6. Details of this result were never published and in the paper mentioned in the heading we shall describe the method in detail, give a complete proof of the necessary results, and perform deferred corrections. As a preview we offer here some numerical results.

<u>Example 5</u>:

$-\Delta u - c^2(\sin cx + \sin cy) = 0$, on the circle of radius 1/2 and center at (0,0).

<u>Exact solution</u> (and boundary function).

$u^*(x,y) = \sin cx + \sin cy$.

For h = 1/28 we obtained the following max. abs. errors for various values of c

Correction	c=1	c=0.1	c=10
0	8.9,-7	9.0,-12	2.5,-2
1	1.2,-10	3.6,-14	2.8,-4
2	2.4,-11	3.4,-14	6.3,-5
3	6.1,-12	4.0,-14	2.9,-5

The high order boundary interpolation is essential. Experiments show that when it is not performed no improvements are obtained with the corrections.

Acknowledgements. Part of this work was prepared when the author was visiting Stanford University during the Winter Quarter of 1973. Computing and other support was provided by Nat. Science Foundation Contract GJ-35135X. My special thanks to Gene H. Golub and Olof Widlund for their constant encouragement.

This work is dedicated to my friend Manuel Bemporad, who through his unceasing activities has promoted Computer Sciences at the Central University to its present position, and has been instrumental in facilitating my research activities there.
The excellent typing is due to Miss B. Cerceau.

REFERENCES

1. Björck, Å. and V. Pereyra, "Solution of Vandermonde systems of equations". Math. Comp. 24, 893-903 (1970).
2. Bulirsch R. and J. Stöer, "Numerical treatment of ordinary differential equations by extrapolation methods". Numer. Math. 8, 1-13 (1966).
3. Ciarlet, P.G., M.H. Schultz, and R.S. Varga, "Numerical methods of high-order accuracy for nonlinear boundary value problems I". Numer. Math. 9, 394-430 (1967).
4. Concus P., and G.H. Golub, "Use of fast direct methods for the efficient numerical solution of nonseparable elliptic equations". Techn. Rep. STAN-CS-72-278, Computer Sciences Dept., Stanford Univ. California (1972).
5. Concus P., G.H. Golub and V. Pereyra, "Fast direct deferred correction solver for mildly nonlinear elliptic equations on rectangular regions". To appear.
6. Crane, P.C. and P.A. Fox, "DESUB-Integration of a first order system of ordinary differential equations". Numerical Math. Comp. Programs 2, Issue 1. Bell Labs., New Jersey (1969).
7. Daniel J.W., V. Pereyra and L.L. Schumaker, "Iterated deferred corrections for initial value problems". Acta Cient. Venezolana 19, 128-135 (1968).
8. Falkenberg, J.C., "A method for integration of unstable systems of ordinary differential equations subject to two-point boundary conditions". BIT 8, 86-103 (1968).
9. Fox, L. (Editor) Numerical Solution of Ordinary and Partial Differential Equations. Pergamon Press, Oxford (1962).

10. Galimberti G. and V. Pereyra, "Numerical differentiation and the solution of multidimensional Vandermonde systems". Math. Comp. 24, 357-364 (1970).

11. ______________________________ , "Solving confluent Vandermonde systems of Hermite type". Numer. Math. 18, 44-60 (1971).

12. Gear, C.W., Numerical Initial Value Problems in Ordinary Differential Equations. Prentice Hall, New Jersey (1971).

13. Henrici, P., Discrete Variable Methods in Ordinary Differential Equations. Wiley, New York (1962).

14. Herbold, R.J., M.H. Schultz, and R.S. Varga, "The effect of quadrature errors in the numerical solution of boundary value problems by variational techniques". aequat. math. 3, 247-270 (1969).

15. Holt, J.F., "Numerical solution of nonlinear two-point boundary problems by finite difference methods". Comm. A.C.M. 7, 366-373 (1964).

16. Jerome, J.W. and R.S. Varga, "Generalizations of spline functions and applications to nonlinear boundary value and eigenvalue problems". In Theory and Applications of Spline Functions, 103-155. Academic Press, New York (1969).

17. Joyçe, C.C., "Survey of extrapolation processes in numerical analysis". SIAM Rev. 13, 435-490 (1971).

18. Kantorovich, L.V. and V.I. Krylov, Approximate Methods of Higher Analysis. Noordhoff, The Netherlands (1958).

19. Keller, H.B., "Accurate difference methods for nonlinear two-point boundary value problems". Manuscript (1972).

20. Keller, H.B., "Accurate difference methods for linear ordinary differential systems subject to linear constraints". SIAM J. Numer. Anal. 6, 8-30 (1969).

21. __________ , Numerical Methods for Two-Point Boundary-Value Problems. Blaisdell, Mass. (1968).

22. Kreiss, H.O. "Difference approximations for ordinary differential equations". Uppsala Univ. Comp. Sc. Dept. Rep. NR 35 (1971).

23. Krogh, F.T., "A variable step variable order multistep method for the numerical solution of ordinary differential equations". Proceeding IFIP-68, 194-199. North-Holland Pub. Co., Amsterdam (1969a).

24. __________ , "VODQ/SVDQ/DVDQ-Variable order integrators for the numerical solution of ordinary differential equations". TU Doc. CP-2308, NPO-11643, Jet Propulsion Lab., Pasadena, Cal. (1969b).

25. Lentini, M., "Correcciones diferidas para problemas de contorno en sistemas de ecuaciones diferenciales ordinarias de primer orden".

Pub. 73-04, Depto. de Comp., Fac. Ciencias, Univ. Central de Venezuela, Caracas (1973).

26. Lentini M. and Pereyra V., "A variable order variable step finite difference method for non-linear multipoint boundary value problems". To appear.

27. Osborne, M.R., "On shooting methods for boundary value problems". J. Math. Anal. App. 27, 417-433 (1969).

28. Pereyra V., "The difference correction method for nonlinear two-point boundary value problems". Stanford Univ. CS18, California (1965).

29. ________, "Accelerating the convergence of discretization algorithms". SIAM J. Numer. Anal. 4, 508-533 (1967a).

30. ________, "Iterated deferred corrections for nonlinear operator equations". Numer. Math. 10, 316-323 (1967b).

31. ________, "Highly accurate discrete methods for nonlinear problems". MRC Techn. Rep. 728, Univ. of Wisconsin, Madison (1967c).

32. ________, "Iterated deferred corrections for nonlinear boundary value problems". Numer. Math. 11, 111-125 (1968).

33. ________, "Highly accurate numerical solution of casilinear elliptic boundary value problems in n dimensions". Math. Comp. 24, 771-783 (1970).

34. ________, "High order finite difference solution of differential equations". Stanford Univ. Report, California (1973).

35. Pereyra V. and G. Scherer, "Efficient computer manipulation of tensor products with application to multidimensional approximation". To appear in Math. Comp. (1973).

36. Pereyra V. and O. Widlund, "A sixth order method for mildly nonlinear elliptic equations on curved boundaries". To appear.

37. Perrin, F.M., H.S. Price, and R.S. Varga, "On high-order numerical methods for nonlinear two-point boundary value problems". Numer. Math. 13, 180-198 (1969).

38. Roberts, S.M., J.S. Shipman, and W.J. Ellis, "A perturbation technique for nonlinear two-point boundary value problems". SIAM J. Numer. Anal. 6, 347-358 (1969).

39. Stetter, H., "Asymptotic expansions for the error of discretization algorithms for nonlinear functional equations". Numer. Math. 7, 18-31 (1965).

40. Widlund, O., "Some recent applications of asymptotic error expansions to finite-difference schemes". Proc. Roy. Soc. London A323, 167-177 (1971).

CYCLIC FINITE-DIFFERENCE METHODS FOR ORDINARY DIFFERENTIAL EQUATIONS

Hans J. Stetter

I. INTRODUCTION

In the solution of certain linear algebraic equations by iterative methods it is common practice to use sequences of parameters in a cyclic fashion in order to reduce error propagation; see, e.g., [6]. In the numerical solution of ordinary differential equations, this approach has only recently been introduced (Donelson and Hansen [3]) although it was involved in a disguised form in several algorithms previously developed. It is the purpose of this talk to review the possible application of this technique and to exhibit its potential and its limitations.

Basically, two objectives may be the goal of using several finite-difference procedures in a cyclic fashion:

(a) improvement of the approximation properties,

(b) improvement of the stability properties.

A trivial example will make this clear. The alternate use of the explicit and implicit EULER-procedures with identical steps h is equivalent to the use of the implicit trapezoidal procedure or the implicit midpoint procedure, with step 2h :

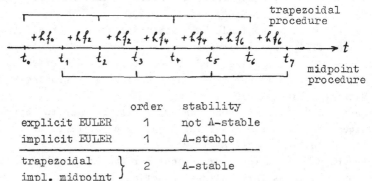

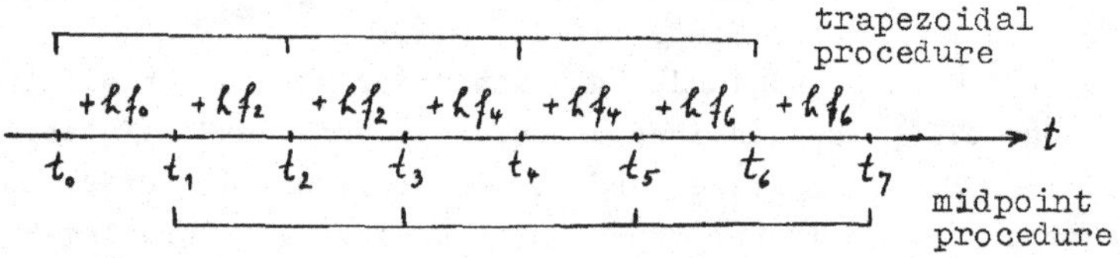

	order	stability
explicit EULER	1	not A-stable
implicit EULER	1	A-stable
trapezoidal / impl. midpoint	2	A-stable

Thus, the "cyclic" methods composed of the explicit and the implicit EULER procedure are more accurate and yet share the A-stability of the implicit EULER method.

Let us now define what we mean by a cyclic discretization method for the solution of an initial value problem

$$(1)\qquad \begin{aligned} y' &= f(t,y),\ t > 0, \\ y(0) &= y_0, \end{aligned}$$

for a system of s ordinary differential equations. For simplicity, we consider only equidistant grids $G_h := \{t_\nu := \nu h,\ \nu = 0,1,\dots\}$ and denote the values of functions $\eta\colon G_h \to \mathbb{R}^s$ by $\eta_\nu := \eta(t_\nu)$. A k-step procedure Φ_h is a relation

$$(2)\qquad \Phi_h(t_\nu;\eta_{\nu-k},\dots,\eta_\nu) = 0$$

which produces an approximation η_ν to the value $y(t_\nu)$ of the solution y of (1) from given approximations $\eta_{\nu-\kappa}$ to $y(t_{\nu-\kappa})$, $\kappa = 1(1)k$.

Definition 1: Consider an ordered set of k_μ-step procedures $\Phi_h^{(\mu)}$, $\mu = 1(1)m$. The finite difference method which consists in using the procedures $\Phi_h^{(\mu)}$ in a cyclic sequence to compute successive values η_ν, is called m-cyclic method[1] based upon $\Phi_h^{(1)},\dots,\Phi_h^{(m)}$.

It is clear that an m-cyclic method is but a special case of a general m-stage method (cf. [5], sect. 5.4). We need only consider the computation of m consecutive values of η as one step of a method which computes the next vector

$$(3)\qquad \boldsymbol{\eta}_{\bar\nu} := \begin{pmatrix} \eta_{\bar\nu m} \\ \eta_{\bar\nu m+1} \\ \vdots \\ \eta_{\bar\nu m+m-1} \end{pmatrix} \in \mathbb{R}^{ms}$$

from one or more previous such vectors. In fact, an emphasis on the cyclic nature of a method will only be justified when the constituent procedures $\Phi_h^{(\mu)}$ have identical structures (e.g., are all linear k-step methods). In the following, we will always assume that this is the case.

[1]) Naturally, a starting procedure must also be specified; we will not specify appropriate starting procedures for our methods, in this report.

II. CYCLIC METHODS WITH INCREASED ORDER

Our first objective will be the construction of cyclic methods which are convergent of an order higher than that which could be attained by a uniform method of the same structure.

We will consider the two classical cases of order restrictions in uniform methods.

In explicit RK-methods it is not possible to attain order m with m stages for $m > 4$ due to the large number of consistency conditions to be satisfied.

In linear k-step methods it is not possible to attain an order of convergence which is equal to the attainable order of consistency for $k > 2$; here the reason is the instability of the highly consistent linear k-step procedures.

1. Explicit m-stage RK methods

Let the individual RK-procedures of our cyclic RK-methods have the same number m of stages and the same order p. The order of the cyclic method will be denoted by q.

Necessary and sufficient for $q = p + 1$ is (see [5], Theorem 5.4.7)

$$\sum_{\mu} (\text{principal error function})_{\mu} = 0 \tag{4}$$

where the sum runs over the RK-procedures of the cycle.

For $p = m$, it is impossible to satisfy (4) with any number of procedures in the cycle since there is one term in the principal error function which is independent of the particular procedures. This excludes an improvement for $m \leq 4$.

For $m = 5$ and 6, the attainable orders p are only 4 and 5 resp. Here it is possible to find pairs of explicit m-stage RK-procedures the alternating use of which yields a method of order $q = m = p + 1$. Such pairs were exhibited by Butcher in a slightly different context. In his paper [2], he obtains 5-stage RK-methods b and c which are of order 3 each; but the composite method bc is of order 5.

Whether there exist cyclic methods with two or more explicit m-stage RK-procedures which achieve an order $q = m$ for $m > 6$ is not known. A positive answer is likely at least for the next few values of m.

If one admits one <u>implicit</u> RK-procedure in the cycle, the attainment of $q > m$ is possible at least for low values of m.

For $m = 1$, we have the explicit and implicit EULER-procedures (cf. sect. I). For $m = 2$, an analogous example with $q = 3$ is

$$\begin{pmatrix} 0 & 0 \\ \frac{2}{3} & 0 \\ \hline \frac{1}{4} & \frac{3}{4} \end{pmatrix} \quad \text{and} \quad \begin{pmatrix} 0 & 0 \\ 0 & \frac{2}{3} \\ \hline \frac{1}{4} & \frac{3}{4} \end{pmatrix}.$$

Similar constructions seem to work for the next few m.

<u>2. Linear k-step methods</u>

There exist one-parameter families of implicit linear k-step procedures which are consistent of order 2k-1. Let $\langle\bar{\rho},\bar{\sigma}\rangle$ define the unique k-step procedure of order 2k and $\langle\bar{\bar{\rho}},\bar{\bar{\sigma}}\rangle$ the unique procedure of order 2k-1 with $\bar{\bar{\rho}}$ a polynomial of degree k-1 and $\bar{\bar{\sigma}}$ of degree k. Then $\langle\bar{\rho} + c\bar{\bar{\rho}}, \bar{\sigma} + c\bar{\bar{\sigma}}\rangle$, with $c \in \mathbb{R}$, defines the desired family. No member of this family is D-stable for $k > 2$. The strategy of [3] combines m of these procedures into a cyclic method which is D-stable and hence convergent of order 2k-1. It turns out that $m = k - 1$ is sufficient.

The meaning of D-stability for cyclic linear multistep methods is obvious. We consider the transition

$$\eta_\nu := \begin{pmatrix} \eta_{\nu-k+1} \\ \vdots \\ \eta_\nu \end{pmatrix} \xrightarrow{\text{m-cycle}} \eta_{\nu+m} := \begin{pmatrix} \eta_{\nu-k+1+m} \\ \vdots \\ \eta_{\nu+m} \end{pmatrix}.$$

For the scalar equation $y' = 0$, this transition is effected by a $k \times k$-matrix Q which depends only on the polynomials ρ_μ of the individual k-step procedures of the cycle. Necessary and sufficient for D-stability is $\|Q^n\| \le S < \infty$ for arbitrary n, or the eigenvalues of Q must be in the closed unit disk and those on the unit circle must possess only linear elementary divisors.

If we arrange the coefficients $\alpha_k^{(\mu)}$ of the ρ_μ into the scheme

$$\left(\begin{array}{cccc|cccc} \alpha_0^{(1)} & \alpha_1^{(1)} & \cdots\cdots & & \alpha_k^{(1)} & & & 0 \\ & \alpha_0^{(2)} & \cdots\cdots & & \alpha_{k-1}^{(2)} & \alpha_k^{(2)} & & \\ 0 & & \ddots & & & & \ddots & \\ & & \alpha_0^{(m)} & \cdots & \cdots\cdots & & & \alpha_k^{(m)} \end{array}\right)$$

$$A_0 \quad \cdots\cdots \quad A_j$$

which we then partition into k×k-matrices A_j, $j = 0(1)J$, $J := [\frac{k-1}{m}] + 1$, from the right (cf. [5], sect. 4.3.1), we find that the eigenvalues of Q are the zeros of the polynomial

(5) $$\rho(x) := x^{k-mJ} \det\left(\sum_{j=0}^{J} A_j x^j\right).$$

Let us normalize all $\langle\rho_\mu, \sigma_\mu\rangle$ such that $\alpha_k^{(\mu)} = 1$; then

(6) $$\rho(x) = x^k + X_1 x^{k-1} + \dots + X_{k-1} x + X_k$$

and the coefficients are the fundamental symmetric functions X_i of the zeros x_i, $i = 1(1)k$, of ρ. On the other hand, (5) implies

(7) $$X_i = \sum_{\Sigma\varkappa_\mu = im} \pm \alpha_{\varkappa_1}^{(1)} \alpha_{\varkappa_2}^{(2)} \dots \alpha_{\varkappa_m}^{(m)}$$

where the $\alpha_\varkappa^{(\mu)}$ are of the form

(8) $$\alpha_\varkappa^{(\mu)} = \bar{\alpha}_\varkappa + c_\mu \bar{\bar{\alpha}}_\varkappa$$

since each $\rho_\mu = \bar{\rho} + c_\mu \bar{\bar{\rho}}$ contains a real parameter c_μ. (7) and (8) show that the X_i are linear functions of the fundamental symmetric functions C_j of the c_μ, $\mu = 1(1)m$:

(9) $$A_{io} + \sum_{j=1}^{m} A_{ij} C_j = X_i, \quad i = 1(1)k.$$

Since all individual ρ_μ as well as ρ possess the principal zero 1, the equations (9) are linearly dependent and the matrix of the A_{ij}, $i = 1(1)k$, $j = 1(1)m$, is only of rank k-1 for $m \geq k-1$. Hence $m = k-1$ is the correct number of procedures in the cycle; in this case one may specify the k-1 extraneous zeros of ρ and compute the C_j from (9). (Note that the A_{ij} are universal numbers for given k and m.)

The remaining step in the proof of the existence of (k-1)-cyclic linear k-step methods convergent of order 2k-1 is now the following. We have to show that we can specify the k-1 extraneous zeros $x_2,\dots,x_k$ of ρ such that (9) defines a set of C_j, $j = 1(1)k-1$, which belong to k-1 <u>real</u> parameters c_μ. Or, more formally:

Let T be the domain in the $\mathbb{R}^{k-1}$ of the C_j which is associated with real c_μ (i.e. the C_j are the coefficients of a (k-1)-degree polynomial with all roots real). Similary let S be the domain in the $\mathbb{R}^{k-1}$ of the symmetric fundamental functions $X_i^!$, $i = 1(1)k-1$, which belong to extraneous zeros x_i, $i = 2(1)k$, inside the open unit disk (i.e. $\rho(x) = (x-1)[x^{k-1} + \sum_{i=1}^{k-1} X_i^! \, x^{k-1-i}]$ satisfies the root criterion). (9) establishes a linear inhomogeneous mapping $\varphi_k : \mathbb{R}^{k-1} \to \mathbb{R}^{k-1}$ from the X-space to the C-space. Necessary and sufficient for the existence of our methods is

$$(10) \qquad \varphi_k(S) \cap T \neq \emptyset.$$

For k = 3, it is not difficult to find that $\varphi_3(S) \subset T$ so that each "stable" choice of the extraneous zeros leads to real parameters c_μ, $\mu = 1,2$. For k = 4, there are examples in [3]. The general proof of (10) in [4] is not quite complete; but it seems that (10) is generally true.

After this surprising success of the cyclic approach one may ask whether it is possible to raise the order q of a cyclic linear k-step method consisting of (2k-1)-order procedures to 2k without loss of stability. The interpretation of the m-cyclic method as an m-stage J-step method permits the application of Theorem 5.4.3 from [5] which implies - under the assumption that all extraneous zeros of ρ are in the open unit disk - that one homogeneous condition in the coefficients of the ρ_μ and σ_μ is necessary and sufficient. By a closer analysis of this condition or by a consideration of the cyclic structure it is seen that this condition is linear in the fundamental symmetric functions $C_1,\dots,C_m$ of the parameters. If we append this homogeneous equation to the system (9) and set m = k, we have a similar situation as previously:
(9) establishes a linear inhomogeneous mapping ψ_k from the $\mathbb{R}^{k-1}$ of the X_i into that (k-1)-dimensional subspace of the $\mathbb{R}^k$ of the C_j, $j = 1(1)k$, which is defined by the order 2k condition. Let $\overline{T}$ be the intersection of T in $\mathbb{R}^k$ with this subspace; then we have to establish

$$(11) \qquad \psi_k(S) \cap \overline{T} \neq \emptyset.$$

The non-emptiness of the intersection has been established for k = 3 and 4 (see [3]); the general proof is still deficient.

For m = k = 2, one obtains the curious result that a 2-cyclic method consisting of the SIMPSON-procedure and an arbitrary implicit 3rd order D-stable linear 2-step procedure is of order 4.

Two more remarks: The analogous construction is possible for explicit linear k-step procedures; here one obtains (k-1)-cyclic methods of order 2k-2 and k-cyclic methods of order 2k-1, at least for $k \leq 4$. Also one may substitute predictor-corrector procedures for pure correctors in the cyclic methods with implicit procedures.

III. CYCLIC METHODS WITH INCREASED STABILITY

In this section, "stability" will mean "absolute stability" (see, e.g., [5], sect. 2.3.6). The stability region $H_o \subset \mathbb{C}$ (more precisely, region of absolute stability) of a method is defined in the usual sense: when the method is applied to $y' = gy$, $g \in \mathbb{C}$, with step h, then the solution of the difference equation decreases exponentially if $hg \in H_o$.

Our objective is to find cyclic methods which have larger stability regions than uniform methods with procedures of the same structure. We start with a simple example.

Let us form an m-cyclic method consisting of m EULER-procedures with different individual steps $b_\mu h$, $\mu = 1(1)m$, $\Sigma b_\mu = m$. The characteristic polynomial - whose zeros define H_o - of such a method is obviously

$$\varphi(x,H) = x - \prod_{\mu=1}^{m} (1+b_\mu H)$$

so that the stability region is the part of the complex H-plane in which the polynomial

$$p(H) = \prod_{\mu=1}^{m} (1+b_\mu H) \tag{12}$$

takes values inside the unit disk.

A reasonable choice for p is

$$p(H) = e\, T_m(\frac{H}{r} + s); \tag{13}$$

T_m is the m-th Chebyshev polynomial, e is some number slightly smaller than 1, and r and s are chosen such that $p(o) = 1$, $p'(o)=m$, which is necessary to match the representations (12) and (13).

In this fashion one obtains regions H_o which are nearly m-times as long along the negative H-axis than the circular stability region with radius 1 of the EULER method.

In order to construct stability regions for cyclic methods based on k-step procedures we need an expression for the characteristic polynomial of such a method. Again the interpretation as an m-stage J-step method which was used to obtain (5) yields the desired result: if we partition the scheme of the coefficients $\beta^{(\mu)}_{\varkappa}$, $\varkappa = 1(1)k$, $\mu = 1(1)m$, into matrices B_j, $j = 1(1)J$, in the same fashion as we did it for the $\alpha^{(\mu)}_{\varkappa}$ in section II.2, we can deduce

$$\varphi(x,H) = x^{k-mJ} \det \left(\sum_{j=o}^{J} (A_j - HB_j)x^j \right)$$

from a general result on multistage methods ([5], Theorem 5.5.1).

As long as we restrict ourselves to the combination of explicit k-step procedures, nothing spectacular can be achieved. Baron [1] has analyzed the case k = 2 and found that a sizeable increase of the stability regions is possible only when the order is kept artificially low. For higher k, the same result will presumably hold.

If we admit implicit linear k-step procedures, the following question becomes interesting (at least from a mathematical point of view). Is it possible with cyclic linear k-step schemes to break the second "Dahlquist barrier", i.e. to overcome the restriction that uniform linear k-step methods have at most order 2 when they are A-stable? Or in other words, is it possible to construct cycles of implicit linear k-step procedures such that the cyclic method is A-stable and has at least order 3?

Although it is obvious that Dahlquist's proof does not hold for cyclic methods (or multistage methods in general), systematic experiments have failed so far to produce full A-stability when order 3 was enforced. (That $A(\alpha)$-stable methods with α close to $\frac{\pi}{2}$ exist for orders greater than 2 has been established even for linear k-step methods.)

It may be a bit artificial, but we may even interpret the use of <u>smoothing</u> as a cyclic procedure. Assume that smoothing is applied automatically every 10 steps; then we do have a 10-cyclic method which consists of 9 identical "normal" procedures and 1 abnormal procedure which is really the previous procedure plus the smoothing procedure.

Of course, the objective of smoothing is the improvement of the error propagation pattern, i.e. the improvement of the stability properties. As in ordinary cyclic methods this improvement is achieved by breaking the propagation cycle of the parasitic components.

The relation between the two concepts is easily seen by formulating GRAGG's symmetric smoothing procedure for the explicit two-step midpoint method

$$\bar{\eta}_n := \frac{1}{2}(\eta_n+\eta_{n-1}+hf(\eta_n))$$

together with the previous step $\eta_n := \eta_{n-2} + 2hf(\eta_{n-1})$ as a 2-step procedure:

$$\bar{\eta}_n := \frac{1}{2}\eta_{n-1} + \frac{1}{2}\eta_{n-2} + \frac{h}{2} f(\eta_{n-2}+2hf(\eta_{n-1})) + hf(\eta_{n-1}).$$

It is well-known that the smoothed midpoint method is no longer weakly stable (see [5], sect. 6.3.2).

Similary, if we construct a "cyclic" method consisting of m applications of the implicit trapezoidal or the implicit midpoint procedure followed by symmetric smoothing, we obtain a method which no longer possesses an unsatisfactory damping for components of the differential equation with extremely large time-constants. At the same time the evenness in h of the asymptotic expansion is preserved so that "quadratic" Richardson extrapolation in still feasible. This method is well applicable even to very stiff systems of ordinary differential equations.

Let us finally remark that one may even exhibit examples where smoothing not only removes weak stability but at the same time raises the order, a typical phenomenon for cyclic methods.

IV. CONCLUSIONS

The preceding review shows that no dramatic results are to be expected from a systematic exploitation of the idea of "cyclicity" in the numerical treatment of initial value problems for ordinary differential equations.

On the other hand it seems worthwhile to understand the various ramifications of this approach, even if it is only to obtain a fuller insight into the mechanisms of existing algorithms.

Finally, from the point of view of pure mathematics, the concept of "cyclicity" provides a new field for gaining aesthetically pleasing (though numerically irrelevant) results.

References

[1] W. Baron: Optimale Stabilitätsgebiete bei Zweischrittverfahren. Thesis, Techn. Univ. Vienna 1972

[2] J. Butcher: The effective order of Runge-Kutta methods. Conf. on the numer. solution of diff. equns., Lecture Notes in Mathematics No. 109, 133-139, Springer 1969.

[3] J. Donelson, E. Hansen: Cyclic composite multistep predictor-corrector methods. SIAM J. Numer. Anal. 8, 137-157 (1971).

[4] R. Mischak: Lineare zyklische Multischrittverfahren hoher Ordnung. Thesis, Tech. Univ. Vienna 1972.

[5] H.J. Stetter: Analysis of discretization methods for ordinary differential equations. Springer 1973.

[6] R. Varga: Matrix iterative analysis. Prentice Hall 1962.

THE DIMENSION OF PIECEWISE POLYNOMIAL SPACES, AND ONE-SIDED APPROXIMATION

Gilbert Strang

ABSTRACT

Two separate problems are discussed. One is a question implicit in the whole theory of piecewise polynomials: suppose we consider the space S of all piecewise polynomials of degree p and of continuity class C^q, say on a given triangulation in the plane. Then what is the dimension of S, and what is a convenient basis for this space? The answer is known in a dozen special cases, but not in general. The second question has arisen in the approximation of variational inequalities, but is of independent interest. We are given a nonnegative function u on a domain Ω, and want to approximate it from below by a nonnegative spline or finite element u_h: $0 \leq u_h \leq u$. We sketch a proof that under this constraint the usual order of approximation is still possible.

This research was supported by the National Science Foundation (P22928).

AMS subject classifications: 35J20, 41A15, 41A25, 65N30.

We want to discuss two quite separate results--or, more accurately, one result and one conjecture. We shall reverse the order given in the title, and begin with the question we can answer.

1. One-sided approximation of nonnegative functions

Suppose we are given a nonnegative function $u(x)$ defined on the interval $0 \leq x \leq 1$. We introduce equally spaced nodes at the points $x_j = jh$, and consider the best known of all spline and finite element spaces S_h; it is composed of the piecewise linear functions which are continuous at the nodes. The simplest approximating function in S_h is the interpolate u_I, which agrees with u at the nodes: $u_I(x_j) = u(x_j)$. The order of approximation achieved by this piecewise linear interpolate is well-known:

$$||u - u_I|| \leq C_0 h^2 ||u''||, \; ||u' - u_I'|| \leq C_1 h ||u''||. \tag{1}$$

The norms can either be defined as the maximum value over the interval--in which case the estimates (and the best constants $C_0 = 1/8$ and $C_1 = 1/2$) follow directly from a Taylor series expansion of u--or they may be L_2 (or even L_p) norms. In each case we assume only that $||u''|| < \infty$; the estimates are completely familiar, and fundamental to modern approximation theory.

We want to change the problem slightly. Suppose the piecewise linear approximation u_h is constrained to be nonnegative (which the interpolate will be, since u is) and also to lie below the given function u:

$$0 \leq u_h \leq u. \tag{2}$$

The interpolate violates this last condition for any convex function, e.g., $u = x^2$. What order of approximation is possible under these constraints?

We shall explain below where this problem arose; the application required a similar theorem for functions $u(x,y)$ of two variables [1].

Here we want to stay with the one-dimensional case, to prove that the basic error bounds remain valid, and also to ask the reader for references; we know of none. Taylor [4] has given an excellent survey of constrained approximation, including questions of uniqueness and characterization of the best approximation under the condition $u^h \leq u$; but the optimal estimate under this one-sided constraint is easy to obtain, just by subtracting $C_0 h^2 ||u''||$ from the interpolate. It is the squeeze imposed by the other condition $u_h \geq 0$ which creates the problem.

THEOREM. Given $u \geq 0$, there exists a piecewise linear u_h which satisfies $0 \leq u_h \leq u$, and achieves the optimal order of approximation:

$$||u - u_h|| \leq c_0 h^2 ||u''||, \qquad ||u' - u_h'|| \leq c_1 h ||u''||. \tag{3}$$

Remark. We have not determined the best constants.

Proof. Consider the set of functions in S_h which satisfy $0 \leq u_h \leq u$. This set is non-empty, because it contains the zero function. Our choice u_h will be any maximal element of this set.

At any node x_j, the value of u_h cannot be increased while the other nodal values are kept fixed (because u_h is maximal). This raises two possibilities:

i) $u_h = u$ at the node x_j, or

ii) at some point ξ in $[x_{j-1}, x_j)$ or $(x_j, x_{j+1}]$, $u_h = u$ and $u_h' = u'$.

In the latter case, with u_h tangent to u at the point ξ, a Taylor expansion about ξ gives

$$|u(x_j) - u_h(x_j)| \leq \frac{h^2}{2} \max |u''|. \tag{4}$$

Thus, $u_I - u_h$ is of order h^2 at the nodes. Since it is linear over each interval, it is everywhere of order h^2, and its derivative is everywhere of order h. Applying the triangle inequality to $u - u_h = u - u_I + u_I - u_h$ (we know from (1) that the estimates hold for $u - u_I$) the theorem follows immediately in the maximum norm.

With L_2 norms, the changes are only technical; the Taylor remainder for $f = u - u_h$, expanded around the point ξ where $f = f' = 0$, gives

$$f(x_j) = \int_{\xi}^{x_j} (x_j - x)\, f''(x)\, dx.$$

Recalling that $f'' = u''$, the Schwarz inequality yields

$$(5) \qquad |u(x_j) - u_h(x_j)|^2 \leq ch^3 \int_{x_{j-1}}^{x_{j+1}} |u''|^2\, dx.$$

This replaces (4), in the final steps of the proof, with a local bound on $u_I - u_h$ and $u_I' - u_h'$. Integration over each subinterval introduces an extra factor h, and then summation produces the inequalities (3) for $u_I - u_h$. The triangle inequality completes the proof as before.

This is not the place to discuss generalizations of the Theorem, to piecewise polynomials of higher degree and to functions of several variables. The L_2 argument becomes much more technical [3] in the latter case, because Sobolev is against us: a function with second derivatives in L_2 may not be differentiable at the individual points, and therefore the tangency at ξ (and the whole introduction of Taylor series) has to be reexamined.

We do want to describe the application of our Theorem (in two variables) to a variational inequality known as the <u>obstacle problem</u>. The problem is to minimize $I(v) = \iint |\text{grad } v|^2$ over the convex set

$$K = \{v \mid v \in \mathcal{H}_0^1(\Omega), \quad v \geq \psi \text{ throughout } \Omega\}.$$

The possibility that the minimizing u may lie on the boundary of K turns the usual condition $\delta I = 0$ into an inequality--just as, for a function f on the interval $0 \leq x \leq 1$, the possibility of minima at the endpoints of the interval alters the usual requirement $f' = 0$. The Fichera-Stampacchia-Lions condition for u to solve the obstacle problem is

(6) $$\iint_{\Omega} \text{grad } u \cdot \text{grad } (v-u) \, dx \, dy \geq 0 \text{ for all } v \text{ in } K.$$

Suppose we approximate K by the set K_h of all continuous piecewise linear functions v_h, on a given triangulation of Ω, which lie above ψ at the nodes: $v_h \geq \psi_I$ in Ω. Then the Ritz-Galerkin approximation u_h is the function which minimizes $I(v_h)$ over the set K_h, and is determined by its own variational inequality:

(7) $$\iint_{\Omega} \text{grad } u_h \cdot \text{grad } (v_h-u_h) \, dx \, dy \geq 0 \text{ for all } v_h \text{ in } K_h.$$

The problem is to estimate the error $u - u_h$. In a joint paper with Mosco [1], we have applied the one-sided approximation theorem to establish the optimal error bound

(8) $$\iint |\text{grad}(u-u_h)|^2 \, dx \, dy \leq C h^2 \iint (\psi^2 + \psi_{xx}^2 + \psi_{xy}^2 + \psi_{yy}^2) \, dx \, dy.$$

2. Piecewise polynomial vector spaces

We begin as before, not by stating our problem in its most general form, but by asking about one comparatively simple special case. The difference is that this time we are unsure of the answer.

Suppose a polygon Ω is triangulated in the usual way--a vertex of one triangle is not permitted to lie part way along an edge of another. Let S be the space of continuously differentiable piecewise cubics: a typical element v takes the form

$$v = a_1 + a_2 x + a_3 y + a_4 x^2 + a_5 xy + a_6 y^2 + \ldots + a_{10} y^3$$

within each triangle, and across any edge both v and its normal derivative v_n are continuous. Our question is: what is the dimension of S, and which elements v form a basis?

If no continuity is required of v, the question becomes trivial: dim S is obviously $10T$ (T = number of triangles). To construct basis functions φ_j, place ten nodes z_i inside each triangle, and let φ_j be the piecewise cubic determined by $\varphi_j(z_i) = \delta_{ij}$.

For continuous (C^0) cubics the nodes have to be placed more carefully--one at each vertex, two more along each edge, and one in the interior of each triangle. Again a basis is determined by $\varphi_j(z_i) = \delta_{ij}$. The continuity of the φ_j between triangles is assured by the four nodes lying along each edge, two at the ends, and two in the interior. These four nodes are common to the two triangles which share the edge, and four values are sufficient to determine a cubic function φ_j along that edge.

To compare the dimension of finite element spaces, we need the observation that vertices, triangles, and edges occur roughly in the ratio 1:2:3. The Euler polyhedron formula gives an exact count: α interior and β boundary vertices produce $2\alpha + \beta - 2$ triangles and $3\alpha + 2\beta - 3$ edges. As the triangulation is refined, β becomes negligible in comparison with α. Therefore the dimension of the C^0 cubic space is asymptotically

$$\dim S = T + V + 2E \sim 9V.$$

This is to be compared with the dimension $10T \sim 20V$ when there are no constraints of continuity.

We cannot expect that a clever placement of the nodes will answer our questions for the C^1 cubics; it is only for certain piecewise polynomial spaces--and these are the ones for which engineers have searched, because they are by far the most convenient in the finite element method--that there will be a simple interpolating basis. Such a basis is defined in general by $D_i\, \varphi_j(z_i) = \delta_{ij}$ [2], where D_i was differentiation of order zero--just function evaluation--in the examples above. The standard cubic splines do not fit this pattern, because their B-spline basis functions are non-zero at three of the nodes. This implies difficulties at the boundaries of a domain, and explains why the cubic Hermite space (C^1 cubics in one variable, a space which does have an interpolating basis) has been preferred by engineers to splines. We shall have to expect that as more continuity is imposed, the support of the basis functions spreads over neighboring elements.

We propose to compute the number of independent C^1 cubics heuristically, as follows. There are ten coefficients in the polynomial over each triangle, or 10T altogether. Across each edge we need four constraints to guarantee continuity, and three more for continuity of the normal derivative (which is a quadratic). This is a total of 7E constraints, but they are not independent. Around any vertex, the quantities v, v_x, and v_y are now certain to be continuous from one triangle to the next. But then continuity between the last triangle and the first, as we circle the vertex, is a redundant constraint. Removing these 3V redundancies from the constraints leaves, as the total number of free parameters,

$$10T - 7E + 3V \sim 20V - 21V + 3V = 2V.$$

This we conjecture to be the (asymptotic) dimension of the space.

Suppose we attempt a similar calculation for the space S_p^q, whose elements are the piecewise polynomials of degree p and continuity C^q. Then there are $(p+1)(p+2)/2$ coefficients in each triangle, and

$$(p+1) + (p) + (p-1) + \dots + (p+1-q)$$

constraints to assure continuity across each edge. As before, there is a redundancy around every vertex for derivatives of order $\leq q$; there are $(q+1)(q+2)/2$ such derivatives. Combining these coefficients, constraints, and redundancies, we are led to the following:

CONJECTURE: For $p \geq 2q$, $\dim S_p^q \sim (p-q)(p-2q)V$.

In the case $q = 0$, the dimension p^2V is correct. I had not expected so neat a formula for general p and q; whether simplicity of the formula lends support to the conjecture is a deep question in metaphysics.

We note that in one variable the corresponding problem is comparatively easy: there are $p + 1$ coefficients in each subinterval, and $q + 1$ constraints at each node, leaving a dimension of $(p - q)$. To establish that this is actually correct in the spline case $q = p - 1$, Schoenberg had to construct the B-splines; it would not be difficult to find a construction for all p and q, but the most convenient choice of basis does not seem to be settled. On a square rather than a triangular mesh in the plane, C^q continuity implies $C^{q,q}$; for example with $q = 1$, the continuity of v_y across edges parallel to the x axis means that $(v_y)_x$ is also continuous. Similarly $(v_x)_y$ is continuous across vertical edges. Therefore, the cross derivative $v_{yx} = v_{xy}$ is everywhere continuous. As a result (we have not written out a detailed proof) the space is a tensor product of one-dimensional spaces, and its dimension is $(p - q)^2 V$.

We note that, according to our conjecture, the quadratic space S_2^1 should be more or less empty. Powell has convinced us, however, that on a special triangulation (a square mesh with all diagonals drawn in) there do exist C^1 quadratics with compact support. There seems to be one extra redundancy of the constraints at the center of each mesh square, which would disappear if the node were shifted away from the center. These nodes make up half the vertices (the others are at the corners of the squares); we conjecture a dimension of $V/2$ for this space.

Because such special triangulations are possible, we want to make our conjecture more precise (and more approachable) in the following way. Let Ω be the unit square, divided into N^2 small squares and then into $2N^2$ triangles by the diagonals of slope $+ 1$. We impose periodicity at the boundaries of Ω, so that each v in S_p^q extends to a 1-periodic function on the whole plane. Then there are exactly N^2 vertices, $2N^2$ triangles, and $3N^2$ edges. <u>On this triangulation we conjecture an exact dimension of</u> $(p-q)(p-2q)N^2$. Furthermore, we believe that <u>there should exist</u> $(p-q)(p-2q)$ <u>functions</u> ψ_i, <u>whose translates</u>

$$\psi_{ijk} = \psi_i(x - j/N, y - k/N), \quad j,k = 1,\ldots,N,$$

form a basis for S_p^q.

With this translation invariance, the problem invites a Fourier transformation. The degree of continuity which is awkward to determine at edges and vertices becomes a question of the decay of the Fourier transforms, whose denominators have the simple form $\xi^\alpha \eta^\beta (\xi+\eta)^\gamma$; we need $\alpha + \gamma > q + 1, \beta + \gamma > q + 1$. But we have not found $(p - q)(p - 2q)$ independent numerators, and the problem is genuinely open.

REFERENCES

1. Mosco, U., and Strang, G., One-sided approximation and variational inequalities, Bull. Amer. Math. Soc., to appear.
2. Strang, G., and Fix, G., An Analysis of the Finite Element Method, Prentice-Hall, Englewood Cliffs (1973).
3. Strang. G., Approximation in the finite element method, Numer. Math. 19, 81-98 (1972).
4. Taylor, G.D., Uniform approximation with side conditions, manuscript for the Conference on Approximation Theory, Austin, Texas, 1973.

THE COMPARATIVE EFFICIENCY OF CERTAIN FINITE ELEMENT AND FINITE DIFFERENCE METHODS FOR A HYPERBOLIC PROBLEM

Blair Swartz* and Burton Wendroff†

1. Introduction

Numerical analysts are ultimately concerned with the efficiency of the computational schemes they devise.

One way to evaluate the relative efficiency of numerical schemes for evolutionary partial differential equations is to carefully program them for a computer and then compare running times with the observed precision; see, e.g., Culham and Varga [1].

A second approach, as observed by Douglas [2], involves the use of the first few terms of the truncation error. That is, the error and the total computational work are assumed given, respectively, by

$$e = c_p(\Delta x)^p + c_q(\Delta t)^q , \qquad W = c_W/(\Delta x\ \Delta t) ,$$

with coefficients here assumed independent of Δx and Δt. For given e, W is minimized. We find $p\,c_p(\Delta x)^p = q\,c_q(\Delta t)^q$ at the minimum. The resulting

$$(1.1) \qquad W_{min} = c_W(c_p/q)^{1/p}\,(c_q/p)^{1/q}\left[(q+p)/e\right]^{1/p+1/q}$$

can be used to compare various schemes. Strang [8] and Walsh [14] explore the use of this notion in the design of difference schemes.

A third method of comparing efficiencies is to devise a characteristic but elementary model problem and explore how well a class of its solutions is approximated by those of the difference schemes; e.g. Thompson [12], Fromm [4], Kreiss and Oliger [7]. The periodic hyperbolic problem

$$(1.2) \qquad \partial u/\partial t = \partial u/\partial x , \quad u(0,t) = u(1,t) , \quad u(x,0) = \exp(2\pi i \omega x) ,$$

(ω integral) has $u = \exp[2\pi i\omega(x+t)]$ as its solution. Most approximating algorithms possess similar solutions. Using the notation

$$v \equiv (v_0, \dots, v_J)^T , \quad v_j \approx u(x_j) , \quad x_j \equiv j/J \equiv jh ,$$

---

*Work supported by the U. S. Atomic Energy Commission.

†Work supported partly by the U. S. Atomic Energy Commission and partly by the National Science Foundation.

the differential-difference approximation to (1.2)

$$dv/dt = Sv\,, \quad v_0(t) = v_J(t)\,, \quad v_j(0) = \exp(2\pi i\omega x_j)$$

typically is satisfied by

$$\left[v_\omega(t)\right]_j \equiv \exp\left[2\pi i\omega(x_j + ct)\right] \quad ;$$

where the velocity $c = c(\omega, h)$ is explicitly computable from the difference operator S. That is, the component of each frequency in the initial data travels undiminished, but at its own speed.

As an example, let $(Sv)_k = (D_0 v)_k \equiv (v_{k+1} - v_{k-1})/(2h)$. Then

$$dv_\omega/dt = (2\pi i\omega c)v_\omega\,, \quad \text{while} \quad Sv_\omega = [i\sin(2\pi\omega h)/h]\,v_\omega \quad .$$

The difference in phase angle between v_ω and u at the mesh points is

$$2\pi(1-c)\omega t = \epsilon \cdot P\,, \qquad P \equiv \omega t = \#\text{ time periods computed} \quad .$$

ϵ, the phase error per time period, is given in this example by

$$\begin{aligned} \epsilon &= 2\pi\left[1 - i\sin(2\pi\omega h)/(2\pi i\omega h)\right] \\ &= 2\pi\left[1 - b(\theta)/(i\theta)\right], \qquad \theta \equiv 2\pi\omega h \quad . \end{aligned}$$

$b(\theta)$ is the amplification factor, or symbol, of hD_0.

Suppose an implicit approximation to $\partial/\partial x$ is used:

$$(\partial/\partial x)v \approx (A^{-1}Bv)/h \quad ,$$

where each row of A is the translate of the previous one and similarly for B:

$$A_{jk} = \alpha_{k-j}\,, \qquad B_{jk} = \beta_{k-j} \quad .$$

If we use the semi-discretization

$$\text{(1.3)} \qquad A\,dv/dt = Bv/h\,, \quad v_0(t) = v_J(t)\,, \quad v(0) = v_\omega(0) \quad ;$$

then the phase error per period is

$$\epsilon = 2\pi\left\{1 - b(\theta)/[i\theta\, a(\theta)]\right\} \quad ,$$

where a and b are the amplification factors of A and B:

$$\text{(1.4)} \qquad a(\theta) \equiv \Sigma\,\alpha_j e^{ij\theta}\,, \qquad b(\theta) \equiv \Sigma\,\beta_j e^{ij\theta} \quad .$$

(The role of rational trigonometric approximations of θ is here clear, as ϵ is simply the relative error in such an approximation.)

We can and shall express this phase error per period in terms of a more picturesque variable, namely the number of intervals per wavelength,

$$N \equiv 1/(\omega h) \quad :$$

$$\epsilon = 2\pi - N b(2\pi/N)/[i\, a(2\pi/N)] \quad . \tag{1.5}$$

For a given ϵ, each competing difference scheme will require a certain N to yield that phase error per period. We may regard those needing smaller N as more efficient (for that error tolerance ϵ).

2. Three semi-discretizations.

We shall compare, in this fashion, three differential-difference schemes (1.3).

The first spatial discretization is the family of high-order explicit (i.e. $A = I$) centered difference schemes presented in Kreiss and Oliger [7]. They will be represented here simply by the amplification factor of B:

$$b_{C-d}(m, \theta) \equiv i \sin\theta \sum_{\nu=0}^{m-1} \left[\nu!\; 2\sin^{\nu}(\theta/2)\right]^2/(2\nu+1)! \quad .$$

Each scheme may be reconstituted from its symbol as follows: with $D_0 v$ as before and with $D_+D_- v \equiv [v(x+h) - 2v(x) + v(x-h)]/h^2$, replace $i\sin\theta$ by D_0 and $\sin^2(\theta/2)$ by $-h^2 D_+D_-/4$. From the fact [6, Section 9.121, relation 14] that

$$\theta/\sin\theta = \text{arc}\,\sin\tau/\left(\tau\sqrt{1-\tau^2}\right) = \sum_{\nu=0}^{\infty} \left(\nu!\; 2\tau^{\nu}\right)^2/(2\nu+1)! \tag{2.1}$$

if $\tau = \sin(\theta/2) < 1$, we may conclude that the relative error in using b_{C-d} as an approximation to $i\theta$ is $O(\theta^{2m})$. Furthermore, the associated phase error (1.5) is

$$\epsilon_{C-d}(m, N) \equiv N\sin(2\pi/N) \sum_{\nu=m}^{\infty} \left[\nu!\; 2\sin^{\nu}(\pi/N)\right]^2/(2\nu+1)! \quad . \tag{2.2}$$

The second approximations to $\partial/\partial x$ we consider are the difference schemes Thomée has associated with the matrices involved in Galerkin's method (in space) using smooth splines (discussed in detail elsewhere in this volume by Thomée [11] and Wendroff [15]). The matrices arise by asking that u be approximated by W, of the form

$$W(x, t) = \sum_{1}^{J} V_j(t)\, \varphi_j(x) \quad ,$$

determined by requiring the error in satisfying (1.2) be $\perp$ to the span of $\{\varphi_k\}_1^J$:

$$\langle W_t - W_x , \varphi_k \rangle = 0 , \qquad 1 \le k \le J \ .$$

The φ_j are selected as $\varphi_j(x) \equiv \varphi[(x - x_j)/h]$, where φ is Schoenberg's basic, smooth B-spline of order m (degree $m-1$) [9]. The small support of the φ_j , centered on x_j , means that

$$A_{ij} \equiv \langle \varphi_i , \varphi_j \rangle / h \equiv \alpha_{j-i} , \quad B_{ij} \equiv \langle \varphi_i , d\varphi_j/dx \rangle \equiv \beta_{j-i}$$

have bandwidth $2m-1$. As $\varphi_i(x_j) \neq \delta_{ij}$, the V_j are not $W(x_j , t)$. Thomeé observed that if, in the resulting semi-discrete scheme, the V_j's were taken to be function values v_j (as in (1.3)), then one could prove phase errors at the x_j whose order of accuracy is twice that of the order of global accuracy of best approximation using the same smooth splines. From the known Fourier transform of φ one may find $a(\theta)$ and $b(\theta)$ (1.4) via the Poisson summation formula. The resulting phase error is

$$(2.3) \qquad \epsilon_{sp}(m, N) = -2\pi N \sum_{-\infty}^{\infty} \nu(\nu + 1/N)^{-2m} / \sum_{-\infty}^{\infty} (\nu + 1/N)^{-2m} > 0 \quad ,$$

which is easily shown [10] to be $O(N^{-2m})$, $N \to \infty$. This order of accuracy is roughly equal to the bandwidth of the matrices (as in central differencing).

The existence of a third class of schemes, implicit central differencing, was recently pointed out to us by Kreiss. Like (2.1) it arises from a trigonometric approximation of $\theta/\sin\theta$; this time from a Padé rather than a Taylor approximate [13, p. 345, p. 380]

$$\arcsin\tau/\left(\tau\sqrt{1-\tau^2}\right) \approx g_m(\tau) \equiv \frac{1}{1} - \frac{1\cdot 2\tau^2}{3} - \frac{1\cdot 2\tau^2}{5} - \cdots - \frac{(2j-1)2j\tau^2}{(4j-1)} - \frac{(2j-1)2j\tau^2}{(4j+1)} - \cdots - \frac{(2m-1)2m\tau^2}{(4m-1)} \ .$$

We then have $b_m(\theta)/a_m(\theta) \equiv i \sin\theta\, g_m[\sin(\theta/2)]$, with relative accuracy $O(\theta^{4m})$ in approximating $i\theta$. The case $m = 1$ coincides with Thomeé's piecewise linear, tridiagonal, $O(h^4)$ scheme

$$(Av)_j = (v_{j+1} + 4v_j + v_{j-1})/6 , \quad (Bv)_j/h = (v_{j+1} - v_{j-1})/2h$$

while the second pentadiagonal scheme, $O(h^8)$ accurate, is

$$A\,v = (1, 16, 36, 16, 1)v/70\ , \qquad B\,v/h = (-5, -32, 0, 32, 5)v/(84\,h) \quad .$$

To compare these three families of schemes we tabulate, in Table I(a)-(c), the number of mesh intervals per wavelength, N, required to attain phase errors per period of 10^{-2} and of 10^{-4}. An integer p in the first column means the order of accuracy of the phase error is N^{-p} for large N.

Table I : Intervals per Wavelength for

(a) Spline-Galerkin Differencing (2.3)

Type Spline (Accuracy)	$\epsilon = 10^{-2}$	$\epsilon = 10^{-4}$
Linear (4)	8.7	27
Quad. (6)	4.8	9.7
Cubic (8)	3.6	6.0
Quartic (10)	3.1	4.5
Quintic (12)	2.9	3.8
Sextic (14)	2.7	3.4

(b) Implicit Centered Differencing

Order of Accuracy	$\epsilon = 10^{-2}$	$\epsilon = 10^{-4}$
4	8.7	27
8	3.9	6.7
12	3.1	4.3
16	2.7	3.5
20	2.6	3.1
24	2.5	2.8

(c) Explicit Central Differencing (2.2)

Order of Difference Formula	$\epsilon = 10^{-2}$	$\epsilon = 10^{-4}$
2	64.4	644
4	13.3	42.5
6	7.9	17.3
8	6.0	11.0
12	4.6	7.0
16	4.0	5.5
20	3.6	4.8

(d) Smooth Spline Interpolation (5.1)

Order of Interpolant	$e = 10^{-2}$	$e = 10^{-3}$	$e = 10^{-4}$
4	5.0	8.4	14.6
6	3.4	4.6	6.4
8	3.0	3.6	4.5
10	2.7	3.1	3.7
12	2.6	2.9	3.3
14	2.5	2.7	3.0

The spline differences appear to be more efficient than explicit central differences. For a given order of accuracy, they also have slightly stronger resolving power than the implicit central differences. However, comparing results for the same bandwidth (same position in the Tables I(a), (b)) the improved resolving power of implicit central differencing is apparent.

3. Full space-time discretization.

We now discretize (1.3) in its time variable.

The trapezoidal rule is appropriate for implicit approximations to $\partial/\partial x$:

$$A(w^{n+1} - w^n)/\Delta t = B(w^{n+1} + w^n)/2h\,, \quad w^0 = v_\omega(0)\,, \; w_j^n \approx u(jh, n\Delta t) \quad .$$

For this scheme we find

$$w^n = (R)^n w^0\,, \quad n = t/\Delta t$$

where, with $M \equiv 1/(\omega\Delta t)$ = number of (time) intervals per period,

$$R = \left[a(\theta) + \pi b(\theta)/(M\theta)\right] / \left[a(\theta) - \pi b(\theta)/(M\theta)\right]$$

$$\equiv \exp(i\,\sigma_{Trap}) \qquad\qquad (\theta \equiv 2\pi\omega h = 2\pi/N) \quad .$$

Since a is real and b is imaginary, σ is real; the scheme is stable. Again different frequencies jump along, undiminished, at their own velocities. We further find, with ϵ given by (1.5), that

$$\tan(\sigma_{Trap}/2) = \xi/2\,, \qquad \xi \equiv (2\pi - \epsilon)/M = \xi(N, M) \quad .$$

The phase difference, per period, between u and w is

(3.1) $\qquad E_{Trap} \equiv 2\pi - M\sigma_{Trap} \quad .$

The fourth-order in time Padé-diagonal scheme,

$$(A^2 - \lambda AB/2 + \lambda^2 B^2/12)w^{n+1} = (A^2 + \lambda AB/2 + \lambda^2 B^2/12)w^n\,, \qquad \lambda \equiv \Delta t/h \quad ,$$

is likewise stable, with pure phase error

(3.2) $\qquad E_{P-4} \equiv 2\pi - M\sigma_4\,, \qquad \tan(\sigma_4/2) = \xi/\left[2(1 - \xi^2/12)\right] \quad .$

The leap-frog scheme (midpoint rule), $w^{n+1} - w^{n-1} = 2\lambda B w^n$, is more appropriate for explicit centered differencing. Assuming stability, (see Fornberg [3]), we have

(3.3) $\qquad E_{L-f} \equiv 2\pi - M\sigma_{L-f}\,, \qquad \sin\sigma_{L-f} = \xi \quad .$

In each case we have expressed the phase error, per wavelength per period, as $E = E(N, M)$. To compare various schemes we let d be the computational work per mesh point in space-time; and, for a given E, we attempt to find the pair (N, M) which minimizes

$$W = dMN = \text{the work per wavelength per period.}$$

We assume d is independent of N and M, as other computations associated with FFT (d = const. $\log_\omega N$, a slowly varying function of N) lead to similar comparisons.

Table II displays some of the optimal pairs (N, M) for splines with the midpoint and Padé discretizations.

The leap-frog error (3.3) is not a monotone decreasing function of M and N. The difficulty stems from the cancellation of the space and time parts of the phase error so that indeed, there may be no error at all (e.g. $\partial/\partial x \approx D_0$, $\Delta t/h = 1$). This non-monotonicity makes the constrained minimization difficult and its results unrealistic. To solve this problem we expand (3.3) in $1/M$, drop high-order terms, and make the rest positive:

$$(3.4) \qquad \hat{E}_{L-f} \equiv \epsilon + (2\pi)^3/(6M^2) \quad .$$

This estimate is identical with the more conventional estimates [10] of $u - w$; we use it in constructing Table III. Similar estimates of (3.1), (3.2) [10] change Table II very little.

Table II : (N, M) for Spline-differencing in Space and

Spline Type (Accuracy)	(a) Trapezoidal time differencing (3.1) $E = 10^{-2}$	(a) $E = 10^{-4}$	(b) 4th-order Padé time differencing (3.2) $E = 10^{-2}$	(b) $E = 10^{-4}$
Linear (4)	11.4 , 55	36 , 557	10.4 , 7.1	32 , 23
Quad. (6)	5.9 , 52	12 , 524	5.5 , 6.7	11.3 , 22
Cubic (8)	4.3 , 50	7.2 , 505	4.1 , 6.5	6.8 , 21
Quartic (10)	3.6 , 49	5.3 , 493	3.5 , 6.4	5.1 , 20.6
Quintic (12)	3.3 , 48.1	4.4 , 485	3.2 , 6.3	4.2 , 20.4
Sextic (14)	3.0 , 47.6	3.9 , 479	2.9 , 6.2	3.7 , 20.2

Table III : (N, M) for Leap-frog in Time (3.4)

(a) Explicit Centered Space Differencing

Accuracy	$\hat{E} = 10^{-2}$	$\hat{E} = 10^{-4}$
2	91 , 91	909 , 909
4	17.6 , 79	56 , 788
6	10.0 , 75	21.8 , 743
8	7.5 , 72	13.5 , 720
12	5.5 , 70	8.3 , 697
16	4.6 , 68.8	6.4 , 685
20	4.2 , 68.1	5.5 , 678

(b) Spline-differencing in Space

Spline Type (Accuracy)	$\hat{E} = 10^{-2}$	$\hat{E} = 10^{-4}$
Linear (4)	11.4 , 79	36 , 787
Quad. (6)	5.9 , 73	12 , 741
Cubic (8)	4.3 , 71	7.2 , 714
Quartic (10)	3.6 , 69	5.3 , 697
Quintic (12)	3.3 , 68	4.4 , 686
Sextic (14)	3.0 , 67.5	3.9 , 678

It is possible to approximate E even more simply [10] as

$$E \approx c_p N^{-p} + c_q M^{-q} \quad .$$

The approach leading to (1.1) then yields estimates of the minimizing pairs (N, M) in the tables; estimates which turn out to be accurate except that the estimates of N in the lower parts of the spline tables are up to 25% low.

Computations were done, but are omitted, for implicit centered differencing in space. The remarks at the end of Section 2 apply.

4. Conclusions.

First we make an observation concerning Tables II and III. The only place in the Tables where the typical stability limit, $M > N$, is equalled or violated is in the two locations where the spatial order of accuracy is as low as the temporal order, i.e., the top of Table III (a) and the top right of Table II (b). This is predictable from the relation above (1.1).

We are certainly attracted by the 4th order time differencing, at least for $E = 10^{-4}$. We could surely invert the pentadiagonal matrix involved in the upper-right corner of Table II (b) in fewer than the $(909)^2/(32 \times 23) \approx 1200$ operations per mesh point which lets it compete with the diamond difference scheme (Table III (a), upper-right). Indeed, this 4th order piecewise linear Padé scheme competes well with most of the $E = 10^{-4}$ column of Table III (a), especially if we contemplate differencing a problem $\partial u/\partial t = \partial(a u)/\partial x + b$ in the form $dv/dt = A^{-1}B(a v)/h + b$, so that the main work in the Padé scheme is a back substitution.

More general conclusions are more difficult, and it might be wiser to let the reader form his own. Nevertheless, suppose it takes most of the computational time to simply evaluate the coefficient of the differential equation at the points of the space-time net. Then we may take $d = 1$ for all schemes, and compare only products MN. Suppose $E = 10^{-2}$. Any leap-frog scheme will have $N > 2$ and $M > [100(2\pi)^3/6]^{1/2} \approx 64$ for an $MN > 128$. This product exceeds all of the products MN in the $E = 10^{-2}$ column of Table II (b); the latter schemes would be judged more efficient.

Indeed, coefficient evaluations and storage are both considerations favoring comparison of schemes with the same N. If we use the FFT to evaluate all difference operators and their inverses, then (in many problems) the work per mesh point is the same for all schemes with the same N. For each E (and same N) it is the ratio of the M values which then measures the relative efficiencies of the types of differencing. For example, the cubic spline-trapezoidal scheme requires $N = 4.3$ for $E = 10^{-2}$. This N is matched by a 20th order centered difference, leap-frog scheme. The ratio of the M values, $68/50 \approx 1.4$, slightly favors the cubic-spline, trapezoidal scheme. The ratio $68/6.5 \approx 10$ similarly compares the cubic spline, Padé scheme with the same 20th order centered-difference, leap-frog scheme. For $E = 10^{-4}$, $685/23 \approx 32$ compares the cubic spline, 4th order Padé scheme with the 16th order centered difference, leap-frog scheme. On this basis we would conclude that the spline, 4th order Padé schemes are best.

We attempt to extrapolate our conclusions to nonlinear problems in [10].

5. The resolving power of smooth spline interpolation.

The reader may wonder how to convert a high-order accurate result at the equally-spaced mesh points into a high order global approximation. One way is to find a trigonometric approximation, using as many frequencies as one has confidence in. Another way is to use a smooth spline interpolant.

Golomb [5] has considered in detail the errors involved in smooth spline interpolation, at uniformly placed knots, of $\exp[2\pi i \omega x]$. From (5.9) and Lemma 6.3 of that paper, if e is the max norm over $[0, 1]$ of the interpolation error using splines of order $2m$ (degree $2m-1$), then

$$(5.1) \qquad e \le 2\gamma/(1+\gamma) \le 4(2/N)^{2m} ,$$

where the number of intervals per wavelength $N > 2$, and where

$$\gamma \equiv \sum_{\nu=-\infty, \nu\neq 0}^{\infty} 1/(1+\nu N)^{2m} .$$

In table I (d) we exhibit the number of intervals per wavelength which, according to the first inequality of (5.1), yield a maximum error of 10^{-2}, 10^{-3}, and 10^{-4}, for orders of accuracy compatible with the orders of accuracy of our other spline differencing tables.

Let s_u and s_w be, respectively, the splines of order $2m$ interpolating the solution $u = \exp[2\pi i \omega(x+t)]$ and its approximation w. Then

$$\|u - s_w\|_\infty \le \|u - s_u\|_\infty + \|s_u - s_w\|_\infty = e + \|s_u - s_w\|_\infty .$$

But $s_u - s_w$ is the spline interpolant, at the knots, of

$$u - w = \left[1 - \exp(-i P E)\right] \cdot \exp\left[2\pi i \omega(x+t)\right] \equiv C u$$

where P is the number of periods computed and E, the phase error of Section 3, is a constant over the spatial mesh. Since $|C| \le PE$,

$$\|s_u - s_w\|_\infty = \|C s_u\|_\infty \le PE(1+e) ;$$

so that

$$\|u - s_w\|_\infty \le e + PE + ePE .$$

Thus, if N and M are picked so that e and PE are less than 10^{-2}, then $\|u - s_w\| \le 2.01 \times 10^{-2}$.

For example, if we compute with a cubic spline trapezoidal scheme for one period desiring an error of $E = 10^{-2}$, we would use $N = 4.3$ (Table II). According to Table I (d), any septic spline (order 8) interpolant of u using more than 3.6 intervals per wavelength satisfies $e \le 10^{-3}$. Hence, in fact, the septic interpolant of w gives a global error less than 1.1×10^{-2}. Indeed, the quintic interpolant of w yields errors better than 2×10^{-2}.

REFERENCES

[1] Culham, W. E., and Varga, R. S., "Numerical methods for time-dependent, nonlinear boundary value problems," Soc. of Petroleum Engineers Journal, v. 11, 1971, pp. 374-388.

[2] Douglas, J., A survey of numerical methods for partial differential equations, in Advances in Computers, v. 2, Academic Press, New York, 1961, pp. 1-54.

[3] Fornberg, B., "On high order approximations of hyperbolic partial differential equations," Dept. of Computer Sciences Rpt. NR 39, Uppsala University, 1972.

[4] Fromm, J. E., Numerical solution of the Navier-Stokes equations at high Reynolds numbers and the problem of discretization of convective derivatives, in Numerical Methods in Fluid Dynamics (J. J. Smolderen, ed.), AGARD Lecture Series, LS-48, Tech. Editing and Reprod., Ltd., Harford House, London, 1972.

[5] Golomb, M., "Approximation by periodic spline interpolants on uniform meshes," J. Approx. Thy., v. 1, 1968, pp. 26-65.

[6] Gradshteyn, I. S., and Ryzhik, I. M., Table of Integrals, Series, and Products; Academic Press, New York, 1965.

[7] Kreiss, H.-O., and Oliger, J., "Comparison of accurate methods for the integration of hyperbolic equations," Tellus, v. 24, 1972, pp. 199-215.

[8] Strang, G., "On the construction and comparison of difference schemes," SIAM J. Numer. Anal., v. 5, 1968, pp. 506-517.

[9] Schoenberg, I. J., "Contributions to the problem of approximation of equidistant data by analytic functions," parts A and B, Quart. Appl. Math., v. 4, 1946, pp. 45-99, pp. 112-141.

[10] Swartz, B., and Wendroff, B., "The relative efficiency of finite difference and finite element methods, I hyperbolic problems and splines"; submitted for publication.

[11] Thomée, V., "Spline-Galerkin methods for initial-value problems with constant coefficients", these notes.

[12] Thompson, P. D., Numerical Weather Analysis and Prediction, Macmillan, New York, 1961.

[13] Wall, H. S., Analytic Theory of Continued Fractions, van Nostrand, New York, 1948.

[14] Walsh, R. J., "Optimization and comparison of partial difference methods, to appear SIAM J. Numer. Anal., 1973.

[15] Wendroff, B., "Spline-Galerkin methods for initial-value problems with variable coefficients," these notes.

SPLINE-GALERKIN METHODS FOR INITIAL-VALUE PROBLEMS WITH CONSTANT COEFFICIENTS

Vidar Thomée

1. Introduction

In this paper we shall analyze in some examples the use of Galerkin's method for the approximate solution of initial-value problems with constant coefficients. The approximate solution will be sought, for fixed time, in a space $\mathcal{S}_h$ of smooth splines based on a uniform mesh with mesh-width h. If the differential equation is $\partial u/\partial t = Pu$ with P a differential operator with respect to x the (continuous time) Galerkin method consists in finding u_h in $\mathcal{S}_h$ such that $\partial u_h/\partial t - Pu_h$ is orthogonal to $\dot{\mathcal{S}}_h$ for all positive t.

In Section 2 we collect some facts about smooth splines of order μ (cf. [7], [8]) which we shall need in the analysis. In particular, we notice that spline interpolation of a smooth function has a global error bound of $O(h^{\mu})$ for small h.

In Sections 3 and 4 we treat the heat equation and a simple first order hyperbolic equation, respectively. In both cases the global error is of the same order as that of the interpolant (cf. e.g. [2], [3], [4], [6], [9], [10]). We shall show, however, that at the mesh-points the Galerkin solution is better than this might lead us to believe; in fact the error at these points are $O(h^{2\mu-2})$ and $O(h^{2\mu})$ for h small, respectively. This phenomenon is sometimes referred to as super-convergence. The proofs will depend on interpreting the Galerkin equations as finite difference equations and then applying known results from the theory of such equations. This approach makes it possible to derive estimates with respect not only to the L_2-norm, which is natural from the formulation of the Galerkin equations, but also with respect to the maximum-norm.

In Section 5, finally, we treat briefly the case of systems in higher dimensions. An example is given of a fourth order parabolic system in two dimensions and two dependent variables for which the Galerkin problem based on multi-splines is not correctly posed. As is well-known, such a system admits stable finite difference approximations of arbitrary order of accuracy (cf. [14]).

The material in Sections 2, 3 and 4 is essentially contained in [11] and [12] and we refer to these papers for details. Generalization to variable coefficients is discussed in [13] and [15] (cf. the paper by Wendroff in this volume). In [11] and [15] is also described how it is possible to analyze by the present methods Galerkin equations obtained by discretizing the differential equation in time, and some numerical experiments are presented.

2. Splines.

Let h be a (small) positive number and let χ be the characteristic function of $[-\frac{1}{2},\frac{1}{2}]$. Set $\varphi = \chi * \ldots * \chi$ (μ factors) and $\varphi_j(x) = \varphi(h^{-1}x-j)$, and consider bounded smooth splines of order μ,

$$\mathcal{S}_h = \{v = \sum_{j \in Z} c_j \varphi_j;\ (c_j)_{-\infty}^{\infty} \text{ bounded}\}.$$

We associate with $\mathcal{S}_h$ the trigonometric polynomial

$$g_\mu(\vartheta) = \sum_{l \in Z} \varphi(l) e^{-il\vartheta}.$$

Using Poisson's summation formula one can easily prove that for $\mu \geq 2$,

$$g_\mu(\vartheta) = \sum_{l \in Z} \hat{\varphi}(\vartheta + 2\pi l) \quad \text{where} \quad \hat{\varphi}(\vartheta) = \hat{\chi}(\vartheta)^\mu = \left(\frac{2\sin\frac{1}{2}\vartheta}{\vartheta}\right)^\mu .$$

In particular, $g_\mu(\vartheta)$ is positive for real ϑ.

Let now v be an element of $\mathcal{C}$, the space of bounded continuous functions. Then since $g_\mu(\vartheta) \neq 0$ there exists exactly one $v_h \in \mathcal{S}_h$ such that

(2.1) $\qquad v_h(lh) = v(lh), \text{ for } l \in Z,$

and the coefficients $(c_j)_{-\infty}^{\infty}$ of v_h are determined by the convolution equation

$$\sum_{j \in Z} c_j \varphi(l-j) = v(lh), \quad l \in Z.$$

The operator S_h defined by $S_h v = v_h$ is bounded on $\mathcal{C}$,

$$\|S_h v\| \leq C\|v\|, \qquad \|v\| = \sup_x |v(x)|,$$

and for $0 \leq s \leq \mu$ there is a C such that

$$\|(S_h - I)v\| \leq Ch^s \|v\|_{\mathcal{C}^s}.$$

We shall also have reason to consider for $\gamma \leq 2\mu-2$ and $\nu = [\frac{1}{2}\sigma]$ the trigonometric polynomial

$$g_{\mu,\sigma}(\vartheta) = h^{\sigma-1}(-i)^{\sigma-2\nu} \sum_{l \in Z} (D^{\sigma-\nu}\varphi_0, D^{\nu}\varphi_l)e^{-il\vartheta}.$$

The factor $h^{\sigma-1}$ makes $g_{\mu,\sigma}$ independent of h and using Poisson's summation formula one can prove

(2.2) $$g_{\mu,\sigma}(\vartheta) = \sum_l (\vartheta+2\pi l)^{\sigma} \hat{\varphi}(\vartheta+2\pi l)^2.$$

In particular, $g_{\mu,0} = g_{2\mu}$.

3. The heat equation.

We shall consider here the initial-value problem for the heat equation,

(3.1) $$\frac{\partial u}{\partial t} = \frac{\partial^2 u}{\partial x^2}, \quad t > 0, \quad -\infty < x < \infty,$$

$$u(x,0) = v(x).$$

The Galerkin method will now consist in finding $u_h(\cdot,t) \in \mathcal{S}_h$ for $t \geq 0$ such that

(3.2) $$(\frac{\partial u_h}{\partial t}, w) + (Du_h, Dw) = 0, \quad \text{for } w \in \mathcal{S}_h \text{ and } t > 0,$$

where $D = d/dx$ and

$$(v,w) = \int_{-\infty}^{\infty} v(x)w(x)dx,$$

and such that $u_h(\cdot,0) = S_h v$. Setting

(3.3) $$u_h(x,t) = \sum_j c_j(t)\varphi_j(x),$$

(3.2) may be written as

$$\sum_j \{c_j'(t)(\varphi_j,\varphi_l) + c_j(t)(D\varphi_j,D\varphi_l)\} = 0 \quad \text{for} \quad l \in Z.$$

Since (φ_j,φ_l) and $(D\varphi_j,D\varphi_l)$ only depend on $l-j$ this is a convolution equation, and introducing the Fourier series

$$\tilde{c}(\vartheta,t) = \sum_j c_j(t)e^{-ij\vartheta},$$

we obtain formally with the notation of Section 2,

$$hg_{\mu,0}(\vartheta)\frac{d}{dt}\tilde{c}(\vartheta,t) + h^{-1}g_{\mu,2}(\vartheta)\tilde{c}(\vartheta,t) = 0.$$

This ordinary differential equation with respect to t may be solved to yield

$$\tilde{c}(\vartheta,t) = \exp\left(-\frac{t}{h^2}\frac{g_{\mu,2}(\vartheta)}{g_{\mu,0}(\vartheta)}\right)\tilde{c}(\vartheta,0).$$

It follows that

$$c_j(t) = \sum_l f_l(t/h^2)c_{j-l}(0),$$

where $f_l(t/h^2)$ are the Fourier coefficients of the exponential,

$$\exp\left(-\frac{t}{h^2}\frac{g_{\mu,2}(\vartheta)}{g_{\mu,0}(\vartheta)}\right) = \sum_l f_l(t/h^2)e^{-il\vartheta}.$$

Using the equation (2.1) for the initial-values one finds that hence

(3.4) $$u_h(jh,t) = \sum_l f_l(t/h^2)v_h(jh-lh) = \sum_l f_l(t/h^2)v(jh-lh).$$

This means that the solution operator $G_h(t)$ of the Galerkin problem at the points of hZ can be thought of as a finite difference operator (with coefficients depending on t/h^2) acting on the initial-values and hence the complete solution operator can be thought of as the composition of a finite difference operator and a spline interpolation operator.

Now let λ be a fixed positive number and introduce an auxiliary variable

k by $k = \lambda h^2$. We shall consider below for convenience t of the form nk where n is a natural number. Let us then introduce the finite difference operator (with constant coefficients)

$$F_k v(x) = \sum_j f_j v(x-jh) \quad \text{where} \quad f_j = f_j(\lambda).$$

This (implicit) finite difference operator has the symbol

$$Q(\vartheta) = \sum_j f_j e^{-ij\vartheta} = \exp\left(-\lambda \frac{g_{\mu,2}(\vartheta)}{g_{\mu,0}(\vartheta)}\right),$$

and the iterated operator F_k^n has the symbol

$$Q(\vartheta)^n = \exp\left(-\frac{t}{h^2} \frac{g_{\mu,2}(\vartheta)}{g_{\mu,0}(\vartheta)}\right) = \sum_j f_j(t/h^2) e^{-ij\vartheta},$$

so that (3.4) may be written

$$u_h(jh, nk) = (F_k^n v)(jh),$$

or

$$u_h(x,t) = G_h(nk)v(x) = S_h F_k^n v(x).$$

We can hence state the following representation result.

<u>Proposition 3.1</u>. For $t = nk$, $k = \lambda h^2$, the solution operator $G_h(t)$ of the Galerkin problem can be represented in the form $G_h(t) = S_h F_k^n$ where F_k is a finite difference operator with symbol

$$Q(\vartheta) = \exp\left(-\lambda \frac{g_{\mu,2}(\vartheta)}{g_{\mu,0}(\vartheta)}\right).$$

The difference operator F_k is parabolic, consistent with (3.1) and accurate of order $2\mu-2$.

<u>Proof</u>. It remains only to prove the statements of the last sentence. One finds easily for $|\vartheta| \leq \pi$,

$$-\lambda \frac{g_{\mu,2}(\vartheta)}{g_{\mu,0}(\vartheta)} = -\lambda\left(2 \sin \tfrac{1}{2}\vartheta\right)^2 \frac{g_{2\mu-2}(\vartheta)}{g_{2\mu}(\vartheta)} \leq -c\lambda\vartheta^2, \qquad c > 0,$$

so that F_k is parabolic (in the sense of F. John). Further, by (2.2), for $\vartheta \to 0$,

$$-\lambda \frac{g_{\mu,2}(\vartheta)}{g_{\mu,0}(\vartheta)} = -\lambda \frac{\vartheta^2 \hat{\chi}(\vartheta)^{2\mu} + O(\vartheta^{2\mu})}{\hat{\chi}(\vartheta)^{2\mu} + O(\vartheta^{2\mu})} = -\lambda\vartheta^2 + O(\vartheta^{2\mu}),$$

so that F_k is consistent with (3.1) and accurate of order $2\mu-2$.

Precise convergence estimates for parabolic difference operators are known, not only in the L_2-norm but also in the maximum-norm (cf. [16]). These may now be applied to yield error estimates at the mesh-points, for instance the following:

Proposition 3.2. For given $T > 0$ there is a C such that for $t = nk \leq T$ and $0 \leq s \leq 2\mu-2$,

$$\|G_h(t)v - S_h u(t)\|_\infty \leq Ch^s \|v\|_{\mathcal{C}^s}.$$

4. A simple hyperbolic equation.

In this section we consider the initial-value problem

$$(4.1) \qquad \frac{\partial u}{\partial t} = \frac{\partial u}{\partial x}, \quad t > 0, \quad -\infty < x < \infty,$$

$$u(x,0) = v(x).$$

This time the Galerkin equation is taken to be

$$(\frac{\partial u_h}{\partial t}, \varphi_l) - (Du_h, \varphi_l) = 0,$$

or, expressed in terms of the coefficients of u_h (cf. (3.3)),

$$\sum_j [c_j'(t)(\varphi_j, \varphi_l) - c_j(t)(D\varphi_j, \varphi_l)] = 0 \quad \text{for} \quad l \in Z.$$

Analogously to above we obtain now for the discrete Fourier transform $\tilde{c}(\vartheta,t)$ the ordinary differential equation

$$hg_{\mu,0}(\vartheta) \frac{d}{dt} \tilde{c}(\vartheta,t) - ig_{\mu,1}(\vartheta)\tilde{c}(\vartheta,t) = 0,$$

or

$$\tilde{c}(\vartheta,t) = \exp(\frac{t}{h} i \frac{g_{\mu,1}(\vartheta)}{g_{\mu,0}(\vartheta)})\tilde{c}(\vartheta,0).$$

Let now λ be a fixed number and introduce this time $k = \lambda h$. Considering $t = nk$ where n is a natural number we have as before:

Proposition 4.1. For $t = nk$, $k = \lambda h$, the solution operator $G_h(t)$ of the Galerkin problem can be represented in the form $G_h(t) = S_h F_k^n$ where F_k is a finite difference operator with symbol

$$\varrho(\vartheta) = \exp(\lambda i \frac{g_{\mu,1}(\vartheta)}{g_{\mu,0}(\vartheta)}).$$

The difference operator F_k is L_2-stable, non-dissipative, consistent with (4.1) and accurate of order 2μ.

Proof. It only remains to show the last statement. Since $g_{\mu,1}$ and $g_{\mu,0}$ are real we have $|\varrho(\vartheta)| \equiv 1$ so that F_k is L_2-stable and non-dissipative. In the same way as above we find

$$g_{\mu,1}(\vartheta) = \vartheta\hat{\chi}(\vartheta)^{2\mu} + O(\vartheta^{2\mu}) \quad \text{as} \quad \vartheta \to 0.$$

This time $g_{\mu,1}$ is an odd function so that we may conclude that in fact

$$\frac{g_{\mu,1}(\vartheta)}{g_{\mu,0}(\vartheta)} = \vartheta + O(\vartheta^{2\mu+1}) \quad \text{as} \quad \vartheta \to 0,$$

which proves that F_k is consistent with (4.1) and accurate of order 2μ.

Since $\varrho(\vartheta) = \exp(i\Phi(\vartheta))$ with ψ real, analytic and non-linear, the operator F_k, although stable in L_2, is not stable with respect to the maximum-norm. For such operators, maximum-norm convergence estimates were nevertheless obtained in [5] and [1]. These may now be applied immediately to yield error estimates at the mesh-points. We may for instance conclude the following result.

Proposition 4.2. For given $T > 0$ there is a C such that for $t = nk \leq T$ and $s \geq 0$, $s \neq \frac{1}{2}(2\mu+1)$, $s \neq 2\mu+1$,

$$\|G_h(t)v - S_h u(t)\|_\infty \leq C h^{\beta(s)} \|v\|_{\mathcal{C}^s}$$

where $\beta(s) = \min(2\mu, s\frac{2\mu}{2\mu+1}, s-\frac{1}{2})$.

As an example, consider the case of piecewise linear functions ($\mu = 2$). Then

$$\varrho(\vartheta) = \exp(i\lambda \frac{\sin\vartheta}{\frac{2}{3}+\frac{1}{3}\cos\vartheta}) = \exp(i\lambda\vartheta + O(\vartheta^5)) \quad \text{as} \quad \vartheta \to 0,$$

so that F_k is accurate of order 4. If $v \in \mathcal{C}^s$ with $s > 5$ the error at the mesh-points is hence $O(h^4)$ as $h \to 0$.

5. A remark on systems.

We shall consider the initial-value problem for the system

$$(D^\alpha = \prod_{j=1}^{d} (\frac{1}{j} \frac{\partial}{\partial x_j})^{\alpha_j})$$

(5.1) $$\frac{\partial u}{\partial t} = P(D)u \equiv \sum_{|\alpha| \leq m} P_\alpha D^\alpha u,$$

where now u is an N-vector and P_α are N×N constant matrices. This initial-value problem is correctly posed in L_2 if and only if

$$|\exp(tP(\xi))| \leq C \quad \text{for} \quad \xi \in R^d, \quad 0 \leq t \leq T.$$

A necessary but not sufficient condition for this is Petrovskiĭ's condition

$$\Lambda(P(\xi)) \leq C, \quad \xi \in R^d,$$

where for a matrix A with eigenvalues λ_j we set $\Lambda(A) = \max_j \operatorname{Re} \lambda_j$, and a sufficient but not necessary condition is semi-boundedness,

(5.2) $$\operatorname{Re}(P(\xi)V,V) \leq c|V|^2,$$

(with obvious notation). In the scalar case these conditions coincide and are necessary and sufficient.

In the same way as in Sections 3 and 4 one can see that with the obvious formulation of the (continuous time) Galerkin problem for (5.1) based on multi-splines of order μ, the solution of this problem at the mesh-points takes the form

(5.3) $$u_h(jh,t) = \sum_{l \in Z^d} f_l(h,t)v(jh-lh), \quad j \in Z^d.$$

Here

$$\sum_{l \in Z^d} f_l(h,t)e^{-i\langle l,\vartheta\rangle} = \exp(t\,\frac{A_h(\vartheta)}{G(\vartheta)}), \quad \langle l,\vartheta\rangle = \sum_j l_j\vartheta_j,$$

where

$$G(\vartheta) = \prod_{j=1}^{d} g_{\mu,0}(\vartheta_j) = \sum_{l \in Z^d} \hat{\Phi}(\vartheta+2\pi l)^2, \quad \hat{\Phi}(\vartheta) = \prod_{j=1}^{d} (\frac{2\sin\frac{1}{2}\vartheta_j}{\vartheta_j})^\mu,$$

$$A_h(\vartheta) = \sum_{l \in Z^d} P(h^{-1}(\vartheta+2\pi l))\hat{\Phi}(\vartheta+2\pi l)^2.$$

We may say that the Galerkin problem is correctly posed in L_2 if the finite difference operator associated with it by (5.3) is uniformly bounded in L_2, or

(5.4) $$|\exp(t\,\frac{A_h(\vartheta)}{G(\vartheta)})| \leq C, \text{ for } \vartheta \in R^d, \quad 0 \leq t \leq T.$$

It is a special case of a result of Swartz and Wendroff [10] that if P is semibounded so that (5.2) holds then the Galerkin problem is correctly posed. In fact,

$$\mathrm{Re}(A_h(\vartheta)V,V) = \sum_l \mathrm{Re}(P(h^{-1}(\vartheta+2\pi l))V,V)\hat{\Phi}(\vartheta+2\pi l)^2 \leq cG(\vartheta)|V|^2,$$

from which (5.4) easily follows. In particular, for scalar equations, correctness of the initial-value problem implies correctness of the corresponding Galerkin problem.

For a first order system, as was noticed by Strang and Fix [9],

$$\frac{A_h(\vartheta)}{G(\vartheta)} = P(\eta) \quad \text{with} \quad \eta_j = \sum_l \frac{(\vartheta_j + 2\pi l_j)\hat{\Phi}(\vartheta+2\pi l)^2}{hG(\vartheta)},$$

so that correctness of the initial-value problem again immediately implies correctness of the Galerkin problem.

We shall show now by an example that the Galerkin problem does not always inherit the correctness properties of the initial-value problem. Consider thus the fourth order system in two space dimensions corresponding to

(5.5) $$P(\xi) = \begin{pmatrix} -|\xi|^4 & \gamma^2\xi_1^4 \\ \gamma^2\xi_2^4 & -|\xi|^4 \end{pmatrix},$$

where γ is a positive parameter to be specified below. We have here

$$\Lambda(P(\xi)) = -|\xi|^4 + \gamma^2\xi_1^2\xi_2^2 \leq -(1-\tfrac{1}{4}\gamma^2)|\xi|^4.$$

Hence, for $\gamma < 2$ the system is parabolic in Petrovskiĭ's sense; in particular it is correctly posed in L_2 and even in the maximum-norm. Let us notice, however, that for $\gamma > \sqrt{2}$, $P(\xi)$ is not semi-bounded. For the largest eigenvalue of the hermitian matrix $\mathrm{Re}\, P(\xi) = \frac{1}{2}(P(\xi) + P(\xi)^*)$ is $-|\xi|^4 + \frac{1}{2}\gamma^2(\xi_1^4 + \xi_2^4)$ and thus choosing $\xi = (\tau, 0)$ we obtain a contradiction to (5.2) for large τ if $\gamma^2 > 2$.

We have the following:

<u>Proposition 5.1</u>. There is a positive γ such that the system defined by (5.5) is parabolic in Petrovskiĭ's sense but such that the corresponding Galerkin problem based on multi-splines of order $\mu \geq 3$ is not correctly posed.

<u>Proof</u>. Since $A_h(\vartheta) = h^{-4}A_1(\vartheta)$ it is sufficient to prove that for some γ and ϑ, $\Lambda(A_1(\vartheta))$ is positive. The elements a_{jk} of $A_1(\vartheta)$ are

$$a_{11} = a_{22} = -\sum_{l} |\vartheta+2\pi l|^4 \hat{\varphi}(\vartheta_1+2\pi l_1)^2 \hat{\varphi}(\vartheta_2+2\pi l_2)^2$$

$$= -g_{\mu,4}(\vartheta_1)g_{\mu,0}(\vartheta_2) - 2g_{\mu,2}(\vartheta_1)g_{\mu,2}(\vartheta_2) - g_{\mu,0}(\vartheta_1)g_{\mu,4}(\vartheta_2),$$

and similarly

$$a_{12} = \gamma^2 g_{\mu,4}(\vartheta_1)g_{\mu,0}(\vartheta_2), \qquad a_{21} = \gamma^2 g_{\mu,0}(\vartheta_1)g_{\mu,4}(\vartheta_2).$$

Notice now that by Schwartz' inequality,

$$g_{\mu,2}(\eta)^2 = \Big(\sum_{l} (\eta+2\pi l)^2 \hat{\varphi}(\eta+2\pi l)^2\Big)^2$$

$$\leq \sum_{l} (\eta+2\pi l)^4 \hat{\varphi}(\eta+2\pi l)^2 \sum_{l} \hat{\varphi}(\eta+2\pi l)^2 = g_{\mu,4}(\eta)g_{\mu,0}(\eta),$$

and that equality holds only for $\eta \equiv 0 \pmod{2\pi}$. Hence fixing η with $0 < |\eta| \leq \pi$ we have

$$g_{\mu,2}(\eta)^2 < g_{\mu,4}(\eta)g_{\mu,0}(\eta). \tag{5.6}$$

A simple calculation gives hence for $\vartheta = (\eta, \eta)$,

$$\Lambda(A_1(\vartheta)) = (\gamma^2-2)g_{\mu,0}(\eta)g_{\mu,4}(\eta) - 2g_{\mu,2}(\eta)^2.$$

For $\gamma = 2$ we conclude from (5.6) that $\Lambda(A_1(\vartheta))$ is positive. Hence this still holds for $\gamma < 2$ sufficiently close to 2. Since it was proved above that the system is parabolic for $\gamma < 2$ this completes the proof.

References

[1] Brenner, Ph., and Thomée, V., Stability and convergence rates in L_p for certain difference schemes. Math. Scand. 27(1970), 5-23.

[2] Douglas, J. Jr., and Dupont, T., Galerkin methods for parabolic equations. SIAM J. Numer. Anal. 7(1970), 575-626.

[3] Dupont, T., Galerkin methods for first order hyperbolics: an example. SIAM J. Numer. Anal. (to appear).

[4] Fix, G., and Nassif, N., On finite element approximations to time dependent problems. Numer. Math. 19(1972), 127-135.

[5] Hedstrom, G.W., The rate of convergence of some difference schemes, SIAM J. Numer. Anal. 5(1968), 363-406.

[6] Price, H.S., and Varga, R.S., Error bounds for semi-discrete Galerkin approximations of parabolic problems with application to petroleum reservoir mechanics. Numerical Solution of Field Problems in Continuum Physics. AMS Providence R.I., 1970, 74-94.

[7] Schoenberg, I.J., Contributions to the problem of approximation of equidistant data by analytic functions, A and B. Quart. Appl. Math. 4(1946), 45-99, 112-141.

[8] Schoenberg, I.J., Cardinal interpolation and spline functions. J. Approximation Theory 2(1969), 335-374.

[9] Strang, G., and Fix, G., A Fourier analysis of the finite element variational method, Mimeographed notes.

[10] Swartz, B., and Wendroff, B., Generalized finite difference schemes, Math. Comp. 23(1969), 37-50.

[11] Thomée, V., Spline approximation and difference schemes for the heat equation. The Mathematical Foundations of the Finite Element Method with Application to Partial Differential Equations. Edited by K. Aziz. Academic Press 1972, 711-746.

[12] Thomée, V., Convergence estimates for semi-discrete Galerkin methods for initial-value problems. Numerische, insbesondere approximationstheoretische Behandlung von Funktionalgleichungen. Springer Lecture Notes, to appear.

[13] Thomée, V., and Wendroff, B., Convergence estimates for Galerkin methods for variable coefficient initial-value problems, to appear.

[14] Wendroff, B., Well-posed problems and stable difference operators, SIAM J. Numer. Anal. 5(1968), 71-82.

[15] Wendroff, B., On finite elements for equations of evolution. Technical Report LA-DC-72-1220, Los Alamos.

[16] Widlund, O.B., On the rate of convergence for parabolic difference schemes, II. Comm. Pure. Appl. Math. 23(1970), 79-96.

ON THE ACCELERATED SSOR METHOD FOR SOLVING ELLIPTIC BOUNDARY VALUE PROBLEMS

David M. Young*

1. INTRODUCTION

The paper is concerned with the use of the symmetric successive overrelaxation method (SSOR method) for solving large systems of linear algebraic equations arising in the solution by finite difference methods of boundary value problems involving elliptic partial differential equations. Our main result is to show that by the use of the SSOR method, combined with an acceleration procedure, one can attain a specified degree of convergence to the exact solution of the linear system in $O(h^{-\frac{1}{2}})$ iterations where h is the mesh size. The result holds for a class of problems involving the self-adjoint equation

$$\frac{\partial}{\partial x}\left(A \frac{\partial u}{\partial x}\right) + \frac{\partial}{\partial y}\left(C \frac{\partial u}{\partial y}\right) + Fu = G \tag{1.1}$$

where

$$A(x,y) > 0, \quad C(x,y) > 0, \quad F(x,y) \leq 0 \tag{1.2}$$

in the region. It is assumed that A(x,y) and C(x,y) belong to class $C^{(2)}$. It is not required that the region be a rectangle. The result represents a considerable improvement over the successive overrelaxation method (SOR method) which requires $O(h^{-1})$ iterations.

2. A MODEL PROBLEM

Let us first consider the following model problem. Given a function g(x,y) defined and continuous on the boundary S of the unit square, find a function u(x,y) continuous in R + S and satisfying Laplace's equation

$$\frac{\partial^2 u}{\partial x^2} + \frac{\partial^2 u}{\partial y^2} = 0 \tag{2.1}$$

in R. Here R is the interior of the square. The function u(x,y) is required to satisfy on S the condition

$$u(x,y) = g(x,y). \tag{2.2}$$

In order to apply the method of finite differences, we choose a positive number h (the mesh size) such that h^{-1} is an integer. At points (ph,qh), where p and q are integers, inside the square we require that the difference equation

$$\frac{u(x+h,y)+u(x-h,y)-2u(x,y)}{h^2} + \frac{u(x,y+h)+u(x,y-h)-2u(x,y)}{h^2} = 0 \tag{2.3}$$

be satisfied. If we require that (2.2) hold for mesh points on S we obtain a system of N linear algebraic equations with N unknowns, where N is the number of mesh points in R.

In the case h = 1/3 and with the points labelled as in Figure 2.1 we have, upon multiplying by $-h^2$,

$$\begin{bmatrix} 4 & -1 & -1 & 0 \\ -1 & 4 & 0 & -1 \\ -1 & 0 & 4 & -1 \\ 0 & -1 & -1 & 4 \end{bmatrix} \begin{bmatrix} u_1 \\ u_2 \\ u_3 \\ u_4 \end{bmatrix} = \begin{bmatrix} g_6 + g_{16} \\ g_7 + g_9 \\ g_{13} + g_{15} \\ g_{10} + g_{12} \end{bmatrix} = \begin{bmatrix} b_1 \\ b_2 \\ b_3 \\ b_4 \end{bmatrix} \tag{2.4}$$

*The author was supported in part by U.S. Army Research Office (Durham) grant DA-ARO-D-31-124-72-G34 at The University of Texas at Austin, U. S. A.

continuation of this contribution see on page 196

ALGEBRAIC-GEOMETRY FOUNDATIONS FOR FINITE-ELEMENT COMPUTATION

Eugene L. Wachspress *

Dedicated to Max Noether on the 100th anniversary of his Fundamental Theorem of Algebraic-Geometry.

1. REFERENCES

Preliminary studies were reported at Dundee in 1970 by Wait(7) and Wachspress (3) and in J.I.M.A. by Wachspress (4,5). A thorough analysis is contained in an as yet unpublished research monograph by Wachspress (6). In 1873, Max Noether (2) announced his Fundamental Theorem describing conditions required of polynomials F,G,H to assure existence of polynomials A and B for which H = AF+BG. His analysis was central to the development of modern algebraic-geometry. Walker's text on Algebraic Curves (8) contains the mathematical background essential for comprehension of this paper.

2. NOTATION AND DEFINITIONS

Capital letters usually denote polynomials over the reals in the projective plane or the curves on which these polynomials vanish. Polynomials differing only in normalization have the same curve and are equivalent. Subscripts denote maximal orders (degrees) of curves (polynomials) and superscripts are identifying indices.

Singularities (multiple points) of curves play a leading role in the analysis. Non-ordinary singularities, where two or more branches have a common tangent, are resolved by quadratic transformations which define neighborhoods of points. If $m_p(F)$ is the multiplicity of F at point p, then the divisor of F on G (or of G on F) is

$$FoG = \sum_p (m_p(F)\, m_p(G))\, p , \tag{1}$$

where the symbolic summation is over all p, including neighbors. It is easily shown that $F^1F^2oG = F^1oG + F^2oG$.

The genus of irreducible curve F_s is the nonnegative number

$$g_s = \tfrac{1}{2}(s-1)(s-2) - \tfrac{1}{2}\sum_i r_i(r_i-1) , \tag{2}$$

where $r_i = m_{p_i}(F_s)$ and the summation is over all singular points, including neighbors, of F_s. A line (s=1) or conic (s=2) has genus zero. An algebraic curve is rational iff it has a rational parametrization. Rationality and genus zero are equivalent. Affine coordinates will be used in this discussion even though the analysis applies in the projective plane.

* Work done under Contract Number W-31-109 Eng-52.

3. ALGEBRAIC-GEOMETRY THEOREMS

A few crucial theorems will now be stated:

Theorem 1 (Bezout). The order of FoG is $O(FoG) = \sum_p m_p(F)m_p(G)$.
If F_t and G_s have no common component, then $O(F_t o G_s) = ts$.

Theorem 2. Let neither P nor R have irreducible curve Q as a component. Iff $PoQ = RoQ$, there is a real b such that polynomial $P - bR$ vanishes everywhere on curve Q. (This is a derivative of Noether's Theorem.)

Theorem 3. Let V be the linear series of algebraic curves of order $\leq (m-3)$ with multiplicities r_i at points p_i. Then the

$$\text{dimension of } V \geq \tfrac{1}{2}m(m-3) - \tfrac{1}{2}\sum_i r_i(r_i+1) \ . \tag{3}$$

Theorem 4. Let P_s have n distinct irreducible components and let all its singular points (including neighbors) be $p_1, p_2, \ldots$ of multiplicities $r_1, r_2, \ldots$ Then

$$\tfrac{1}{2}\, s(s-3) + n - \sum_{j=1}^{n} g_j - \tfrac{1}{2}\sum_i r_i(r_i-1) = 0 \ , \tag{4}$$

where g_s is the genus of component P^j of P_s .

4. ALGEBRAIC ELEMENTS AND PATCHWORK FUNCTIONS

Let curve C_m have n distinct irreducible components: $P^1, P^2, \ldots, P^n$. Let points v_1 in $P^n o P^1$, v_2 in $P^1 o P^2, \ldots, v_n$ in $P^{n-1} o P^n$ be designated as vertices. Let $\bar{P}^i$ denote a given segment of curve P^i between vertices v_i and v_{i+1} (with $v_{n+1} = v_1$) such that the n segments define a simple closed planar figure. This figure is an algebraic element or polypol, and more specifically it is an n-pol of degree m. The polypol is said to be well-set iff

a) the vertices are all ordinary double-points of C_m,

b) the (open) segments $\bar{P}^i$ contain only simple points of C_m, and

c) the polypol interior contains no point of C_m.

Otherwise, the polypol is said to be ill-set.

Let u_q $(q=1,2,\ldots,M)$ be the values of $u(x,y)$ at designated nodes (x_q, y_q), and let $W_q(x,y)$ be associated wedge functions, to be defined. The W_q are a basis for degree k approximation over a well-set polypol iff a) for $u(x,y)$ any polynomial of maximal degree less than k+1: $u(x,y) = \sum_{q=1}^{M} u_q W_q(x,y)$ for all (x,y) in the polypol, and

b) this equation is not true for $u(x,y)$ equal to at least one polynomial of degree k+1.

A planar region is polinized by specification of a collection of well-set polypols which cover the region with no overlap. Patchwork functions, U, are defined over a polinized region by specification of values, U_q, at all polypol nodes and of the associated wedge functions $W_q^j(x,y)$ on each polypol, j:

$$U(x,y) = \sum_{\substack{\text{all } q \\ \text{on element } j}} U_q W_q^j(x,y) \qquad \text{for } (x,y) \text{ in polypol } j .$$

Functions W_q^j and U are not necessarily polynomials. For most polypols they are rational functions.

A patchwork function is continuous over the polinized region and of degree k within each polypol when the wedges are constructed according to the recipe which will now be given.

5. DEGREE ONE APPROXIMATION

A. Node Selection.

All vertices are nodes. If $t > 1$, a node is introduced on segment $\bar{P}_t^i$. This side node is chosen off line (v_i, v_{i+1}). The values of a linear form at any two points determine its value at any point on the line joining these points. A linear form has three degrees of freedom on any curve of order higher than one. Values of a linear form at any three non-collinear points on a curve determine the linear form. Function values at vertices v_i, v_{i+1} and (for $t > 1$) at the side node on $\bar{P}_t^i$ determine a unique linear form iff the three points are not collinear.

B. Terminology.

A side node lies on its adjacent side, and the remaining n-1 sides are opposite. Vertex node v_i has adjacent sides $\bar{P}^{i-1}$ and $\bar{P}^i$ (with $\bar{P}^0 = \bar{P}^n$) and the other (n-2) sides are opposite v_i. The wedge associated with node q is of the form

$$W_q(x,y) = k_q \frac{F^q(x,y)\, R^q(x,y)}{Q_{m-3}(x,y)} \tag{5}$$

The denominator, common to all wedges for a given polypol, is called the adjoint of the polypol. F^q is the opposite factor, and R^q is the adjacent factor. The real number k_q is chosen to normalize W_q to unity at (x_q, y_q).

C. Construction of W_q for Rational Well-Set Polypols.

A polypol is rational iff each of its boundary components is rational. Let boundary curve C_m of a well-set rational n-pol have singular points (including neighbors) p_i of multiplicities $r_i = r(p_i) = m_{p_i}(C_m)$. The definition of well-set assures $r(v_j) = 2$ for $j = 1, 2, \ldots, n$.

Let V^1 be the linear series of curves of maximal order m-3 having multiplicity $r_i - 1$ at each p_i which is not a vertex. Theorems 3 and 4 yield dim $V^1 \geq 0$. Let Q^1 and Q^2 be any two elements in V^1. Then

$$Q^1_{m-3} \circ C_m = Q^2_{m-3} \circ C_m = \sum_{\substack{\text{all singular} \\ \text{points of } C_m}} r_i(r_i-1)p_i - \sum_{j=1}^{n} v_j \,.$$

By Theorem 2, there is a real b such that $Q^1 - bQ^2 = 0$ on C_m. By Theorem 1, $Q^1 - bQ^2$ must contain each irreducible component of C_m. This is possible only if $Q^1 - bQ^2$ is the zero polynomial. Thus, the dimension of V^1 is zero. The unique curve in V^1 is the adjoint of the polypol. This definition of Q_{m-3} is extended to ill-set polypols by allowing for vertices which may not be ordinary double-points of C_m. One condition is removed from the multiplicity requirements on Q_{m-3} for each vertex.

Opposite factor F^q is the product of the factors of polynomial C_m which vanish on the sides opposite node q. The adjacent factor construction is described first for side nodes and then for vertex nodes. Let q be on side P^i_t. There is a unique R^q for which

$$F^q_{m-t} R^q \circ P^i_t = (v_i, v_{i+1}) Q_{m-3} \circ P^i_t \,. \tag{6}$$

By Theorem 1, R^q must be of order t-2. From (6) and the definitions of Q_{m-3} and F^q,

$$R^q_{t-2} \circ P^i_t = (v_i, v_{i+1}) \circ P^i_t - v_i - v_{i+1} + \sum_j s_j(s_j-1)p_j, \tag{7}$$

where $m_{p_j}(P^i_t) = s_j$. Curve P^i_t has zero genus, and we obtain from (2): $\sum_j s_j(s_j-1) = (t-1)(t-2)$. The order of the right hand side of (7) is thus seen to be $(t-2) + (t-1)(t-2) = t(t-2)$, as it must be. Let V^2 be the linear series of curves of maximal order t-2 which have this divisor on P^i_t. Then $m_{p_j}(R^q) = s_j - 1$ and by Theorem 3

$$\dim V^2 \geq \tfrac{1}{2}(t-2)(t+1) - (t-2) - \tfrac{1}{2}\sum_j s_j(s_j-1) = 0.$$

If $R^{q,1}$ and $R^{q,2}$ are any two elements of V^2, then by Thm. 2: $R^{q,1} - bR^{q,2} = 0$ on P^i_t. Since P^i_t is irreducible, this is possible only for $R^{q,1} - bR^{q,2}$ equal to the zero polynomial. Uniqueness of R^q_{t-2} satisfying (6) is thus established.

Application of Thm. 2 to (6) gives for some real number, c,

$$W_q = k_q \frac{F^q R^q}{Q_{m-3}} = c\,(v_i, v_{i+1}) \text{ on } P^i. \tag{8}$$

The side node is not on (v_i, v_{i+1}) so that k_q may be chosen to normalize W_q to unity at (x_q, y_q).

Let q be the node at vertex i and let j_1, j_2 be the side nodes on $\bar{P}^{i-1}$, $\bar{P}^i$, respectively. Let $m_{f_j}(P^{i-1}_{t_1}) = s_j$ and $m_{g_j}(P^i_{t_2}) = z_j$.

Adjacent factor R^q is constructed so that

$$F^q_{m-t_1-t_2} R^q \circ P^{i-1}_{t_1} = (v_{i-1},j_1)\, Q_{m-3} \circ P^{i-1}_{t_1}\ , \text{ and}$$

$$F^q_{m-t_1-t_2} R^q \circ P^{i}_{t_2} = (v_{i+1},j_2)\, Q_{m-3} \circ P^{i}_{t_2}\ . \tag{9}$$

These conditions are equivalent to

$$R^q \circ P^{i-1}_{t_1} = (v_{i-1},j_1)P^i \circ P^{i-1} - (v_{i-1}+v_i) + \sum_j s_j(s_j-1)f_j \text{ when } t_1 > 1,$$

and

$$R^q \circ P^{i}_{t_2} = (v_{i+1},j_2)P^{i-1} \circ P^{i} - (v_{i+1}+v_i) + \sum_j z_j(z_j-1)g_j \text{ when } t_2 > 1. \tag{10}$$

Existence of a unique $R^q_{t_1+t_2-2}$ which satisfies (10) is established with the aid of the algebraic-geometry theorems. Eqtns. 9 yield

$W_q = c_1(v_{i-1},j_1)$ on P^{i-1} and $c_2(v_{i+1},j_2)$ on P^i for some real c_1 and c_2. Vertex v_i lies on neither (v_{i-1},j_1) nor (v_{i+1},j_2), and W_q may therefore be normalized to unity at v_i.

D. Non-Rational Well-Set Polypols

Let g_i be the genus of curve P^i. In the polypol adjoint analysis, dimension $V^1 \geq \sum_{i=1}^{n} g_i$, and to construct a Q_{m-3} one chooses a set of deficit points d^i_j and multiplicities r^i_j so that

for each i: $$\tfrac{1}{2} \sum_j r^i_j(r^i_j-1) = g_i\ . \tag{11}$$

These deficit points are chosen distinct from the singular points of C_m and from the intersections with C_m of the various lines appearing in the adjacent factor analysis. Requiring $m_{d^i_j}(Q_{m-3}) \geq r^i_j-1$ for each i,j yields a unique adjoint.

For each i,j an additional $(r^i_j-1)^2$ elements (not specified) appear in $Q_{m-3} \circ P^i$. These associated points on P^i together with the deficit points replace the "missing" singular points on P^i when $g_i \neq 0$. Unique wedges may be constructed for any prescribed set of deficit points. For a restricted class of deficit point sets, the constructed adjoints do not vanish over the polypol. It is necessary that the deficit and associated points fall on polypol side extensions and not on the $\bar{P}^i$ segments. Henceforth, only rational polypols will be considered in order not to complicate this exposition.

E. Wedge Properties

The rational wedge functions have been constructed so that: 1. $W_q(x_p,y_p) = \begin{matrix} 1 \text{ for } q=p \\ 0 \text{ for } q\neq p \end{matrix}$, 2. $W_q=0$ on sides opposite q, and

3. W_q is linear on side(s) P^i (and P^{i-1}) adjacent to q.

The following crucial property must also be established:

4. W_q is regular over the polypol. That is, $Q_{m-3} \neq 0$ on the polypol.

The construction guarantees $Q_{m-3}\neq 0$ on boundary C_m. All $m(m-3)$ elements of $Q_{m-3}\circ C_m$ are accounted for at multiple points of C_m other than the vertices. An inductive proof of regularity based on the following theorem has been given , Wachspress (6), for a wide class of well-set rational polypols:

Theorem 5. Let $C^1 = P^j P^h P^i$, $C^2 = P^k P^h P^j$, $C^3 = P^k P^h P^i$ be boundaries of well-set polypols T^1, T^2, T^3 such that region T^3 is the union of regions T^1 and T^2. Each of P^i, P^h, P^j, P^k is a product of the irreducible components of the polypol sides, and some of these curves may reduce to points in specific cases, in which case corresponding forms are taken to be unity. Let Q^s be the adjoint of T^s for $s=1,2,3$. If Q^1 and Q^3 are both positive over T^2, then Q^2 is also positive over T^2.

Application to convex polygons illustrates the potency of this theorem. For any triangle, T^1, the adjoint is $Q^1=1$. All triangle wedges are regular. Suppose it has been shown that wedges for all convex (n-1)-gons are regular. Let T^3 be a convex (n-1)-gon obtained by removing factor P^2 from boundary $P^1P^2\ldots P^n$ of convex n-gon, T^2. Then T^1 in Theorem 5 is the triangle bounded by $P^1P^2P^3$ (in the projective plane if P^1 and P^3 are parallel) and the theorem establishes that $Q^2>0$ over T^2. By induction, regularity of wedges for all convex polygons is thus proved. Although similar arguments have been found for all well-set polypols thus far considered, a general proof of regularity has not yet been found.

F. Verification of Continuous Degree One Approximation

Let $u(x,y)$ be linear with nodal values u_i. Over polypol j:

$$u(x,y) - \sum_{\text{all } i \text{ on } j} u_i W_i^j(x,y) \equiv \frac{P_{m-2}(x,y)}{Q_{m-3}(x,y)} \tag{12}$$

, and P_{m-2} vanishes on the polypol boundary as a consequence of the wedge properties. This boundary is of order m and is a product of distinct irreducible curves. Hence, P_{m-2} must be the zero polynomial. Degree one approximation is thus verified. Continuity of the patchwork function is assured by wedge linearity on each polypol boundary component. The constructed wedges form a minimal basis (fewer nodes will not suffice) for continuous degree one patchwork approximation over any polinized region.

6. HIGHER DEGREE APPROXIMATION

One must introduce more nodes to achieve higher degree approximation. Although the precise placement of nodes is arbitrary, there are restrictions similar to those described for degree-one side nodes. The opposite and adjoint polynomial constructions remain unchanged as the degree of approximation is increased. The degrees of adjacent factors increase to yield numerators of degree m+k-3 for degree k approximation over polypols of degree m. For any u(x,y) of maximal degree k, wedges are constructed so that

$$u(x,y) - \sum_{i \atop \text{(all nodes)}} u_i W_i(x,y) \equiv \frac{N_{m+k-3}(x,y)}{Q_{m-3}(x,y)} \qquad (13)$$

vanishes on the boundary, C_m, of the polypol. For $k < 3$, this is possible only when N_{m+k-3} is the zero polynomial. For $k \geq 3$, there is a G_{k-3} such that

$$N_{m+k-3}(x,y) = G_{k-3}(x,y)\; C_m(x,y) \;. \qquad (14)$$

Interior nodes are introduced to yield degree k approximation when $k \geq 3$. Since any $\frac{1}{2}k(k-3)$ points can be located on a curve of order not greater than k-3, one chooses $\frac{1}{2}k(k-3)+1 = \frac{1}{2}(k-1)(k-2)$ interior points which do not all lie on any curve of order $\leq (k-3)$.

The wedge associated with interior node j is

$$W_j = k_j \frac{C_m R^j_{k-3}}{Q_{m-3}} \;, \qquad (15)$$

where R^j_{k-3} is the unique curve of order k-3 which contains all interior nodes but j. The adjacent factor in the numerator of each wedge associated with a boundary node is constructed to vanish at all interior nodes. This assures $N_{m+k-3}(x,y) = 0$ on C_m and also on a set of points contained in no curve of maximal order k-3. Thus, N_{m+k-3} must be the zero polynomial, and degree k approximation is achieved. The wedge associated with boundary node q is

$$W_q = k_q \frac{F^q R^q}{Q_{m-3}} \;. \qquad (16)$$

The numerator factor, F^q, is the product of the forms which vanish on all sides opposite node q.

Suppose q is a side node on P^i_t. Let d(t,k) be the number of degrees of freedom of a polynomial of degree k on a curve of order t. Side nodes are chosen so that any H_k determined by all but one of d(t,k)-2 side nodes and the two vertices on P^i does not contain the deleted side node. This assures unique determination of a polynomial of degree k in terms of its d(t,k) nodal values on P^i. Let the curve obtained by deleting side node q be denoted by H^q_k. Then $O(H^q_k \circ P^i_t) = kt$, and all kt elements are determined by the d(t,k)-1 nodes on the side.

The adjacent factor is constructed so that there is a real b for which $W_q = b\, H_k^q$ on side P^i. By Thm. 2, this is true iff R^q is the curve of maximal order t+k-3 for which
$$Q_{m-3} H_k^q \circ P_t^i = F_{m-t}^q R^q \circ P_t^i . \tag{17}$$
If $r_1, r_2, \ldots$ are the multiplicities of all the singular points $p_1, p_2, \ldots$ (including neighbors) of P^i, then by construction:
$$Q_{m-3} \circ P_t^i = F_{m-t}^q \circ P_t^i - (v_{i-1} + v_i) + \sum_j r_j (r_j - 1) p_j . \quad \text{Hence,}$$
$R^q \circ P_t^i \supseteq \sum_j r_j (r_j - 1) p_j$. This is accomplished if $m_{p_j}(R^q) \geq r_j - 1$, and this imposes at most $\frac{1}{2} \sum_j r_j (r_j - 1)$ conditions on R^q. For P^i rational, this is $\frac{1}{2}(t-1)(t-2)$ conditions. Another tk-2 conditions on R^q yield $R^q \circ P_t^i \supseteq H_k^q \circ P_t^i - (v_{i-1} + v_i)$. The requirement that $R^q = 0$ at all interior nodes imposes another $\frac{1}{2}(k-1)(k-2)$ conditions. Let V^3 be the space of all plane curves of maximal order t+k-3 which satisfy these conditions. Then
$$\text{dimension of } V^3 \geq \tfrac{1}{2}(t+k-3)(t+k) - \tfrac{1}{2}(t-1)(t-2) - \tfrac{1}{2}(k-1)(k-2) - (tk-2) = 0.$$
There is at least one R^q in V^3. Uniqueness is easily demonstrated. The right hand side of $R^q \circ P_t^i = Q_{m-3} H_k^q \circ P_t^i - F_{m-t}^q \circ P_t^i$ does not depend on choice of R^q in V^3. For any two elements of space V^3: $R^{q,1} \circ P_t^i = R^{q,2} \circ P_t^i$, and by Thm. 2 there is a real b for which $R^{q,1} - bR^{q,2} = 0$ on P_t^i. By Thm. 1, there is a polynomial P_{k-3} such that $R^{q,1} - bR^{q,2} = P_{k-3} P_t^i$. Both $R^{q,1}$ and $R^{q,2}$ vanish on the interior nodes (at which $P_t^i \neq 0$) constructed not to all lie on any curve of order $<k-2$. Hence, P_{k-3} must be the zero polynomial. The dimension of V^3 is zero. The adjacent factor is unique. This analysis is reminiscent of the argument which recurs in one-variable Chebyshev minimax theory.

The adjacent factor for node q equal to vertex v_i will now be examined. Let H_k^1 and H_k^2 be the unique curves thru the nodes other than v_i on sides $\bar{P}_{t_1}^{i-1}$ and $\bar{P}_{t_2}^i$, respectively. Factor R^q is chosen so that
$$F_{m-t_1-t_2}^q R^q \circ P_{t_1}^{i-1} = Q_{m-3} H_k^1 \circ P_{t_1}^{i-1} \quad \text{and} \quad F_{m-t_1-t_2}^q R^q \circ P_{t_2}^i = Q_{m-3} H_k^2 \circ P_{t_2}^i .$$
Polynomial R^q is of maximal degree $t_1 + t_2 + k - 3$. Conditions imposed on R^q are similar to those discussed for a side node:

$\frac{1}{2}(t_1-1)(t_1-2) + \frac{1}{2}(t_2-1)(t_2-2)$ for all singular points of P^{i-1} and P^i,

$k(t_1+t_2)-2$ for elements $H_k^1 \circ P_{t_1}^{i-1} - v_{i-1} + H_k^2 \circ P_{t_2}^i - v_{i+1}$,

$\frac{1}{2}(k-1)(k-2)$ for interior nodes, and an additional $t_1 t_2 - 1$ conditions for $P_{t_1}^{i-1} \circ P_{t_2}^i - v_i$. If V^4 is the space of plane curves of maximal order (t_1+t_2+k-3) which satisfy these conditions, then
$$\text{dimension of } V^4 \geq \tfrac{1}{2}(t_1+t_2+k-3)(t_1+t_2+k) - \tfrac{1}{2}(t_1-1)(t_1-2) - \tfrac{1}{2}(t_2-1)(t_2-2)$$
$$-\tfrac{1}{2}(k-1)(k-2) \; - k(t_1+t_2) + 2 - t_1 t_2 + 1 = 0 .$$

If $R^{q,1}$ and $R^{q,2}$ are any two elements of V^4, then $R^{q,1} o P^{i-1}_{t_1} P^i_{t_2} = R^{q,2} o P^{i-1}_{t_1} P^i_{t_2}$ and Thm. 2 assures the existence of a real b for which $R^{q,1} - bR^{q,2} = 0$ on $P^{i-1}_{t_1} P^i_{t_2}$. This is possible iff $R^{q,1} - bR^{q,2} = G_{k-3} P^{i-1}_{t_1} P^i_{t_2}$. Both $R^{q,1}$ and $R^{q,2}$ vanish on the interior nodes, chosen not to all lie on any curve of order $<(k-2)$. Hence, G_{k-3} is the zero polynomial, dimension $V^4 = 0$, and $R^q_{t_1+t_2+k-3}$ is unique.

The conditions imposed on the divisors of R^q on sides adjacent to node q ensure degree k variation of W_q on these sides. The number and placement of side nodes guarantees continuity of the patchwork approximation across polypol boundaries.

7. THREE-DIMENSIONAL ELEMENTS

A three-dimensional algebraic element is called a polypoldron. Let S_{m_2} be an algebraic surface with simple irreducible components $P^1_{s_1}, P^2_{s_2}, \ldots, P^n_{s_n}$. Let section $\bar{P}^i$ of surface P^i be a specified face of an element. Faces intersect on edges which are segments of space curves, and the edges intersect at points which are the polypoldron vertices. If a simple polypoldron has n faces, e edges and v vertices, then Euler's theorem is that $v-e+n=2$. The element is said to be well-set iff it is simple and 1) Each vertex is an ordinary triple-point of S_m, 2) Each edge, excluding the vertices, contains only double-points of S_m, and 3) The polypoldron interior contains no point of S_m. It is easily shown that for any well-set polypoldron $v=2(n-2)$ and $e=3(n-2)$. These relationships are important for wedge analysis. The concept of rational polypoldra is less useful than that of rational polypols. For each edge at the intersection of faces of order t and s one must introduce $ts(\frac{t+s}{2} - 2) + 1$ auxiliary conditions on the adjoint surface and on related adjacent surfaces. Only for special cases can this be accomplished without deficit points. Auxiliary conditions are required on some of the curves of intersection of polypoldra surfaces which are not adjacent. These exterior intersection curves must be contained in the adjoint surface. Wedge construction for a general class of elements is described in reference(6). Space limitations preclude more than a cursory discussion here, and for this reason only convex polyhedra will be considered.

Let $p_1, p_2, \ldots$ be all the non-vertex multiple points, including neighbors, on S_m of multiplicities $r_1, r_2, \ldots$, where all the r_i are greater than two. Then there is a unique adjoint surface, Q_{m-4}, for which $m_{p_i}(Q_{m-4}) \geq r_i - 2$. This surface contains all exterior edges of the polyhedron. In fact, a generalization of Desargues' Theorem establishes a geometric relationship among the multiple-points of S_m which reduces by one the number of conditions imposed by the multiple-points p_i on each exterior curve. This will be clarified later. For a convex polyhedron of degree m, the wedge at vertex node q is of the form:

$$W_q(x,y,z) = k_q \frac{F^q_{m-3}(x,y,z)\; R^q_{k-1}(x,y,z)}{Q_{m-4}(x,y,z)} \quad . \tag{18}$$

For degree k approximation, the numerator is of degree m-4+k. Node placement on edges and faces follows the two-dimensional pattern. For any u of maximal degree k,

$$u(x,y,z) - \sum_i u_i W_i(x,y,z) = \frac{P_{m-4+k}}{Q_{m-4}} \tag{19}$$

vanishes on S_m. Interior points are not required until k=4. The symbol ${}_nC_r$ denotes the binomial coefficient $\binom{n}{r}$. When $k > 3$, one chooses ${}_{k-1}C_3$ interior points which do not all lie on any surface of order less than k-3. The opposite factor, F^q, is the product of the polynomials which vanish on the surfaces opposite node q. Surface R^q is uniquely determined by side, face and interior nodes. For a vertex node, there are: ${}_{k-1}C_1 = k-1$ nodes on each adjacent edge,
${}_{k-1}C_2$ nodes on each adjacent face, and
${}_{k-1}C_3$ nodes in the polyhedron interior.

(The generalization to higher dimensions is apparent.)

A surface of order t has ${}_{t+3}C_3 - 1$ degrees of freedom. The total number of nodes which must be contained in adjacent factor R^q is $3(k-1) + 3\,{}_{k-1}C_2 + {}_{k-1}C_3 = {}_{k+2}C_3 - 1$. There is at least one R^q_{k-1} which contains these nodes. It can be shown that there is only one. For an edge node, the opposite factor is of order m-2 and the adjacent factor is determined by the ${}_{k-2}C_1 = k-2$ adjacent edge nodes,
$2\,{}_{k-1}C_2$ adjacent face nodes, and
${}_{k-1}C_3$ interior nodes.

This is a total of ${}_{k+1}C_3 - 1$ nodes, and determines a unique adjacent surface of order k-2. For a face node, the opposite surface is of order m-1 and the adjacent surface is determined by ${}_{k-1}C_2 - 1$ adjacent face nodes and ${}_{k-1}C_3$ interior nodes, or a total of ${}_kC_3 - 1$ nodes. These yield a unique adjacent surface of order k-3. For an interior node, the opposite surface is of order m and the adjacent surface of order k-4 is uniquely determined by the other ${}_{k-1}C_3 - 1$ interior nodes.

The role of Desargues' Theorem is illustrated by wedge construction for a triangular prism with non-parallel end planes. Here, m=5 and the adjoint is of order m-4=1. This plane is determined by any three non-collinear points. The boundary surface, S_5, has ${}_5C_3 = 10$ triple-points, only six of which are vertices. The other four points must lie on the adjoint plane. Desargues' Theorem is that three of these four points lie on a line. This is the line of intersection of the end planes of the prism. There is a unique plane through this line and the fourth triple-point. In general, to determine the adjoint **surface** one may discard one triple-point on each exterior edge. For the hexahedron described by Wait (7), the adjoint of order two is determined by nine points. There are twelve triple-points in S_6 when the eight vertices are excluded. In general, four of these points lie on each of three exterior edges. The adjoint is constructed as the unique quadric thru the three exterior edges. (Discarding one point on each edge leaves nine points, and a quadric has nine degrees of freedom.) The analysis of degrees of freedom and geometric dependences of multiple-points for the general polypoldron is given in (6) .

8. FURTHER ANALYSIS

A novel application of Hadamard's method of descents (1) enables construction of irrational basis functions for ill-set elements by projecting three-dimensional rational functions on to appropriate surfaces. When considering finite-element approximations to solutions of boundary value problems with partial differential equations of orders higher than two, one often seeks patchwork functions having higher order continuity. Basis functions have yet to be discovered for higher than C^0 continuity over a general polinized region. For such problems, one may formulate the Ritz-Galerkin procedure so that multiple-C^0 function spaces can be used.

Numerical quadrature formulas have been developed to facilitate use of rational wedges over polinized regions in finite-element computation. The versatility of triangles, parallelograms, and isoparametric coordinates is such that there is little need for more general elements to solve problems of practical importance. Nevertheless, it is not unreasonable to envision an increase in sophistication which will require the more general algebraic elements.

9. BIBLIOGRAPHY

(1) Hadamard, J., Lectures on Cauchy's Problem in Linear Partial Differential Equations, Dover, New York (1952).

(2) Noether, M., "Uber einen Satz aus der Theorie der Algerbraischen Funktionen," Math. Ann., 34 (1873), Pp. 447-9.

(3) Wachspress, E., "A Rational Basis for Function Approximation," Proc. of Conf. on Applic. of Numerical Analysis, Dundee (1971), Springer Verlag Lecture Notes in Mathematics, 228 , Pp. 223-252.

(4) Ibid. J. Inst. Math. Applics., 8 (1971), Pp. 57-68.

(5) Wachspress, E., "A Rational Basis for Function Approximation. Part II: Curved Sides," J. Inst. Math. Applics., 11, (1973) Pp. 83-104.

(6) Wachspress, E., A Rational Finite-Element Basis, (1973) Unpublished Monograph.

(7) Wait, R., "A Finite-Element for Three-Dimensional Function Approximation," (same as (3)), Pp. 348-352.

(8) Walker, R., Algebraic Curves, Dover, New York (1962).

SPLINE-GALERKIN METHODS FOR INITIAL-VALUE PROBLEMS WITH VARIABLE COEFFICIENTS

Burton Wendroff*

In his work on initial value problems with constant coefficients Thomeé [1] has shown that the spline-Galerkin method generates a nodal scheme which has a higher convergence rate than approximation theory would indicate. Thomeé and I have extended this to variable coefficients in [2] and [3]. My lecture today is a simplified, slightly reorganized version of this extension.

The problem is

$$\frac{\partial u}{\partial t} = \sum_{\alpha=0}^{m} p_\alpha(x,t) D^\alpha u \equiv P(x,t,D)u \quad , \tag{1}$$

with 1-periodic initial data. The function u could be a vector and the p_α matrices, but for simplicity we will suppose the problem is scalar. The operator P is assumed to be semi-bounded; for $0 \le t \le T$,

$$\int_0^1 P(x,t,D)u \cdot u\,dx \le c(T) \int_0^1 u^2\,dx \quad ,$$

for real 1-periodic functions u.

The basis functions will span the space of 1-periodic splines of order μ based on a regular mesh with spacing $h = 1/N$. Let $\Phi_1, \cdots, \Phi_N$ be such a basis. The Galerkin procedure generates two matrices

$$A = h^{-1}\left\{\int_0^1 \Phi_k \Phi_j\,dx\right\} \quad , \tag{2}$$

and

$$B = h^{-1}\left\{\int_0^1 (P\Phi_k)\Phi_j\,dx\right\} = B(P) \quad , \tag{3}$$

and then defines an approximate solution

$$w(x,t) = \sum v_j(t)\,\Phi_j(x) \quad ,$$

*Research supported by NSF Grant No. G.P. 30305

by

$$(4) \qquad A\,\frac{dv}{dt} = B\,v\ ,$$

$$v(0) \text{ specified ,}$$

where

$$v(t) = \left(v_1(t), \ldots, v_N(t)\right)^T .$$

Thomeé has shown that if the $p_\alpha(x, t)$ are constant then we may compare v directly with u. Indeed, if

$$v(0) = u(0)\ ,$$

and

$$\|v - u\|_h^2 = h \sum_{i=1}^{N} |v_i(t) - u(x_i, t)|^2\ ,$$

then

$$(5) \qquad \|v - u\|_h = O(h^\nu)\ ,$$

where

$$\nu = 2\mu - m + \delta\ ,$$

$$\delta = \delta(m) = \begin{cases} 0, & m \text{ even} \\ 1, & m \text{ odd} \end{cases}$$

This means that it is advantageous to interpret (4) as a differential-difference or nodal scheme, so that the matrices A and B become difference operators.

To obtain (5) in general, note that the error $e = \{e_i\} = \{u(x_i, t) - v_i(t)\}$ satisfies

$$A\,\frac{de}{dt} = B\,e + F\ ,$$

where the truncation error F is

$$F = A\,\frac{\partial u}{\partial t} - B\,u = A\,P(x, t, D)u - B\,u\ .$$

It follows easily from the semi-boundedness of P and the stability of the basis that (5) holds if $F = O(h^\nu)$, and this part of the argument is omitted.

In trying to determine F it is first of all clear that the time t appears only as a parameter which can be suppressed. Second, it is also clear that it is sufficient to show that

$$F_\alpha \equiv A p_\alpha D^\alpha u - B(p_\alpha D^\alpha)u \ ,$$

$$= O\left(h^{2\mu - \alpha + \delta(\alpha)}\right) , \quad 0 \le \alpha \le m \ ,$$

for then $F = \sum_{\alpha=0}^{m} F_\alpha = O(h^\nu)$. Finally, since u is assumed to be a smooth periodic function it is enough to show that for

$$P = p(x)\, D^\alpha \ ,$$

$$\nu' = 2\mu - \alpha + \delta(\alpha) \ ,$$

that

$$f \equiv |A p(x)\, D^\alpha e^{2\pi i \omega x} - B e^{2\pi i \omega x}| = O(h^{\nu'}) \ .$$

Turning to an examination of A and B, set

$$\Phi_j = \sum_{\ell \in z} \varphi_{j + \ell N} \ ,$$

where

$$\varphi_r(x) = \varphi\left(\frac{x}{h} - r\right) \ ,$$

and φ is the B-spline of order μ. Suppose N is even. There is no loss in generality in evaluating f at $x = 1/2$. Then for N sufficiently large,

$$A_\ell \equiv A_{N/2,\, N/2+\ell} = \int_{-\infty}^{\infty} \varphi(s - \ell)\, \varphi(s)\, ds$$

$$B_\ell \equiv B_{N/2,\, N/2+\ell} = \int_{-\infty}^{\infty} \varphi^{(\alpha)}(s - \ell)\, p(1/2 + sh)\, \varphi(s)\, ds$$

and

$$f = |\Sigma A_\ell p(1/2 + \ell h)(2\pi i \omega)^\alpha e^{i\ell\theta} - \Sigma B_\ell e^{i\ell\theta}|$$

for $\theta = 2\pi\omega h$. The coefficient $p(x)$ is effectively replaced by its Taylor expansion at $x = 1/2$ if we set $p(x) = (x - 1/2)^r$, $0 \le r \le r_0$, r_0 to be determined. Then

$$f = |\Sigma A_\ell \ell^r h^r (2\pi i \omega)^\alpha e^{i\ell\theta} - \Sigma B_\ell e^{i\ell\theta}|$$

$$= |h^r (2\pi i \omega)^\alpha i^{-r} D_\theta^r \Sigma A_\ell e^{i\ell\theta} - \Sigma B_\ell e^{i\ell\theta}| .$$

Putting

$$\hat{\varphi}(\xi) = \int_{-\infty}^{\infty} e^{-i\xi s} \varphi(s)\, ds ,$$

it is by now classical that

$$\sum B_\ell e^{i\ell\theta} = i^{\alpha-r} h^{r-\alpha} \sum_{j=-\infty}^{\infty} \hat{\varphi}(\theta + 2\pi j)(\theta + 2\pi j)^\alpha \hat{\varphi}^{(r)}(\theta + 2\pi j)$$

$$\equiv h^{r-\alpha} i^{\alpha-r} g_0(\theta) ,$$

and

$$\sum A_\ell e^{i\ell\theta} = \sum_{j=-\infty}^{\infty} |\hat{\varphi}(\theta + 2\pi j)|^2$$

$$\equiv g_1(\theta) ;$$

so that we must show that

$$|g_1^{(r)}(\theta) - \theta^{-\alpha} g_0(\theta)| = O(\theta^{\nu'-r}) , \tag{6}$$

for small θ, $0 \le r \le r_0 = \nu - 1$.

A basic property of splines is that $\hat{\varphi}(\theta + 2\pi j)$ has a zero of order μ if $j \ne 0$. Then

$$g_1^{(r)}(\theta) = D_\theta^r \left(\hat{\varphi}(\theta)\right)^2 + O(\theta^{2\mu-r}) .$$

For α even,

$$\theta^{-\alpha} g_0(\theta) = \hat{\varphi}(\theta)\,\hat{\varphi}^{(r)}(\theta) + o(\theta^{\nu'-r}) \quad .$$

That this is true for α odd follows from the fact that

$$\hat{\varphi}(\theta + 2\pi j)\,\hat{\varphi}^{(r)}(\theta + 2\pi j) - \hat{\varphi}(\theta - 2\pi j)\,\hat{\varphi}^{(r)}(\theta - 2\pi j)$$

has a zero of order $2\mu - r + 1$ for $j > 0$.

To finally establish (6) it must be shown that

(7) $$D_\theta^r\,|\hat{\varphi}(\theta)|^2 - \hat{\varphi}(\theta)\,\hat{\varphi}^{(r)}(\theta) = o(\theta^{\nu'-r}) \quad ,$$

$0 \le r \le \nu' - 1$. This establishes the constant coefficient case at once, since then only $r = 0$ is needed. Unfortunately, (7) is not true for $r > 0$ if

$$\hat{\varphi}(\theta) = \left(\frac{\theta}{2}\right)^{-\mu} \left(\sin\frac{\theta}{2}\right)^{\mu} \quad .$$

We can construct another spline $\psi(s)$ using

$$\psi(s) = \Sigma c_j\,\varphi(s - j) \quad .$$

Now if

$$\hat{\psi}(\theta) = 1 + o(\theta^{2\mu}) \quad ,$$

then

$$D_\theta^r \left(\hat{\psi}(\theta)\right)^2 - \hat{\psi}(\theta)\,\hat{\psi}^{(r)}(\theta) = o(\theta^{\nu'-r}) \quad .$$

Since

$$\hat{\psi}(\theta) = \Sigma c_j\, e^{ij\theta}\,\hat{\varphi}(\theta)$$

$$\equiv c(\theta)\,\hat{\varphi}(\theta) \quad ,$$

this means finding $c(\theta)$ such that

(8) $$c(\theta)\,\hat{\varphi}(\theta) = 1 + o(\theta^{2\mu}) \quad ,$$

and, for stability, $c(\theta) \neq 0$, $-\pi \le \theta \le \pi$.

Suppose the c_j can be found. Let

$$C = \{c_{kj}\}, \quad C_{kj} = c_{j-k} \quad .$$

If we now formulate the <u>original</u> problem (1) in terms of the new basis ψ we would have

$$C^* A C \frac{d\tilde{v}}{dt} = C^* B C \tilde{v} \quad .$$

Putting

$$\tilde{v}_i(0) = u(x_i, 0) \quad ,$$

the analysis now shows that

$$\|\tilde{v} - u\|_h = O(h^\nu) \quad .$$

To actually obtain $\tilde{v}(t)$ we need only solve

$$A \frac{dv}{dt} = B v \quad ,$$

$$v(0) = C u(0) \quad ,$$

and then set

$$\tilde{v}(t) = C^{-1} v(t) \quad .$$

The existence of the trigonometric polynomial $c(\theta)$ is shown in [2]. In particular if $\mu = 2$, $m = 1$, then $\varphi(s)$ is piecewise linear and $c_0 = 7/6$, $c_1 = c_{-1} = -1/12$.

We have required that each $p_\alpha(x)$ have a local Taylor expansion with remainder $O(h^{\nu'})$. Since $\nu' \geq \nu$, it is sufficient that the remainder be $O(h^\nu)$.

The difference equation

$$C^* A C \frac{\tilde{V}^{n+1} - \tilde{V}^n}{\Delta t} = C^* B C \frac{1}{2}\left(\tilde{V}^{n+1} + \tilde{V}^n\right)$$

has the local truncation error $O(h^\nu) + (\Delta t^2)$. A semiboundedness argument shows that

$$\|\tilde{V}^n - u(n\Delta t)\|_h = O(h^\nu) + O(\Delta t^2)$$

and, again, $\tilde{V}$ is obtained by solving

$$A \frac{V^{n+1} - V^n}{\Delta t} = B \frac{1}{2}\left(V^{n+1} + V^n\right)$$

$$V^0 = C u(0)$$

and putting

$$\tilde{V}^n = C^{-1} V^n \quad .$$

REFERENCES

[1] Thomeé, V., Convergence estimates for semi-discrete Galerkin methods for initial value problems, Numerische, insbesondre approximations-theoretische, Behandung von Functionalgleichungen, Springer Lecture Notes, to appear.

[2] Thomeé, V., and Wendroff, B., "Convergence estimates for Galerkin methods for variable coefficient initial-value problems," to appear SIAM J. Numer. Anal.

[3] Wendroff, B., "On finite elements for equations of evolution," Los Alamos Scientific Laboratory Report No. LA-DC-72-1220, 1972.

This can be written in the matrix form

$$Au = b \tag{2.5}$$

and also in the alternative form

$$u = Bu + c \tag{2.6}$$

or

$$\begin{bmatrix} u_1 \\ u_2 \\ u_3 \\ u_4 \end{bmatrix} = \begin{bmatrix} 0 & 1/4 & 1/4 & 0 \\ 1/4 & 0 & 0 & 1/4 \\ 1/4 & 0 & 0 & 1/4 \\ 0 & 1/4 & 1/4 & 0 \end{bmatrix} \begin{bmatrix} u_1 \\ u_2 \\ u_3 \\ u_4 \end{bmatrix} + \begin{bmatrix} c_1 \\ c_2 \\ c_3 \\ c_4 \end{bmatrix} \tag{2.7}$$

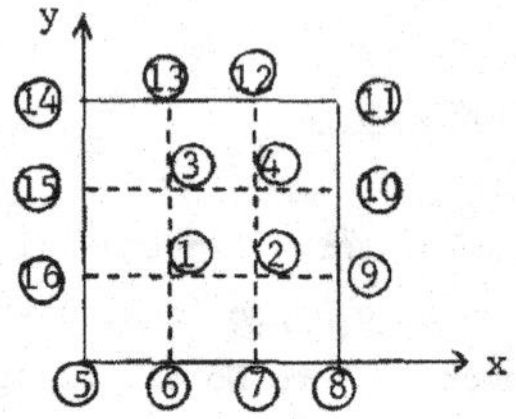

Figure 2.1 - A Model Problem

We also define L and U as strictly lower and strictly upper triangular matrices, respectively, such that

$$L + U = B. \tag{2.8}$$

In the present case we have

$$L = \begin{bmatrix} 0 & 0 & 0 & 0 \\ 1/4 & 0 & 0 & 0 \\ 1/4 & 0 & 0 & 0 \\ 0 & 1/4 & 1/4 & 0 \end{bmatrix}, \quad U = \begin{bmatrix} 0 & 1/4 & 1/4 & 0 \\ 0 & 0 & 0 & 1/4 \\ 0 & 0 & 0 & 1/4 \\ 0 & 0 & 0 & 0 \end{bmatrix} \tag{2.9}$$

We now define three basic iterative methods. The first, the <u>Jacobi method</u>, is defined by

$$\begin{cases} u_1^{(n+1)} = \frac{1}{4}u_2^{(n)} + \frac{1}{4}u_3^{(n)} + c_1 \\ u_2^{(n+1)} = \frac{1}{4}u_1^{(n)} + \frac{1}{4}u_4^{(n)} + c_2 \\ u_3^{(n+1)} = \frac{1}{4}u_1^{(n)} + \frac{1}{4}u_4^{(n)} + c_3 \\ u_4^{(n+1)} = \frac{1}{4}u_2^{(n)} + \frac{1}{4}u_3^{(n)} + c_4 \end{cases} \tag{2.10}$$

or

$$u^{(n+1)} = Bu^{(n)} + c. \tag{2.11}$$

The <u>Gauss-Seidel method</u> is defined by

$$\begin{cases} u_1^{(n+1)} = \frac{1}{4}u_2^{(n)} + \frac{1}{4}u_3^{(n)} + c_1 \\ u_2^{(n+1)} = \frac{1}{4}u_1^{(n+1)} + \frac{1}{4}u_4^{(n)} + c_2 \\ u_3^{(n+1)} = \frac{1}{4}u_1^{(n+1)} + \frac{1}{4}u_4^{(n)} + c_3 \\ u_4^{(n+1)} = \frac{1}{4}u_2^{(n+1)} + \frac{1}{4}u_3^{(n+1)} + c_4 \end{cases} \tag{2.12}$$

or

(2.13) $$u^{(n+1)} = Lu^{(n+1)} + Uu^{(n)} + c.$$

This is equivalent to

(2.14) $$u^{(n+1)} = \mathcal{L}u^{(n)} + (I-L)^{-1}c$$

where

(2.15) $$\mathcal{L} = (I-L)^{-1}U.$$

The successive overrelaxation method (SOR method) is defined by

(2.16) $$\begin{cases} u_1^{(n+1)} = \omega\left\{ \frac{1}{4}u_2^{(n)} + \frac{1}{4}u_3^{(n)} + c_1\right\} + (1-\omega)u_1^{(n)} \\ u_2^{(n+1)} = \omega\left\{\frac{1}{4}u_1^{(n+1)} + \frac{1}{4}u_4^{(n)} + c_2\right\} + (1-\omega)u_2^{(n)} \\ u_3^{(n+1)} = \omega\left\{\frac{1}{4}u_1^{(n+1)} + \frac{1}{4}u_4^{(n)} + c_3\right\} + (1-\omega)u_3^{(n)} \\ u_4^{(n+1)} = \omega\left\{ \frac{1}{4}u_2^{(n+1)} + \frac{1}{4}u_3^{(n+1)} + c_4\right\} + (1-\omega)u_4^{(n)} \end{cases}$$

This is equivalent to

(2.17) $$u^{(n+1)} = \omega(Lu^{(n+1)} + Uu^{(n)} + c) + (1-\omega)u^{(n)}$$

or

(2.18) $$u^{(n+1)} = \mathcal{L}_\omega u^{(n)} + (I-\omega L)^{-1}\omega c$$

where

(2.19) $$\mathcal{L}_\omega = (I-\omega L)^{-1}(\omega U + (1-\omega)I).$$

The rapidity of the Jacobi, Gauss-Seidel, and SOR methods depends on the spectral radii of the matrices B, $\mathcal{L}$, and $\mathcal{L}_\omega$, respectively. Roughly speaking, the error is reduced by a factor of the spectral radius on each iteration. The number of iterations of a method with matrix G is proportional to the reciprocal rate of convergence

(2.20) $$RR(G) = -\frac{1}{\log S(G)}.$$

The following results are well known for the model problem (see, for instance, Young [19]).

Method	Spectral Radius	Reciprocal Rate of Convergence
Jacobi	$S(B) = \cos \pi h$	$RR(B) \doteq \frac{2}{\pi^2} h^{-2}$
Gauss-Seidel	$S(\mathcal{L}) = \cos^2 \pi h$	$RR(\mathcal{L}) \doteq \frac{1}{\pi^2} h^{-1}$
SOR	$S(\mathcal{L}_{\omega_b}) = \frac{1-\sin \pi h}{1+\sin \pi h}$	$RR(\mathcal{L}_{\omega_b}) \doteq \frac{1}{2\pi} h^{-1}$

The value of the optimum relaxation factor ω_b is given by

(2.21) $$\omega_b = \frac{2}{1+\sqrt{1-S(B)^2}} = \frac{2}{1+\sin \pi h}$$

It can be seen that the SOR method converges faster by an order-of-magnitude than the other two methods. We shall later show that an additional order-of-magnitude can be obtained using the SSOR method combined with acceleration.

3. MORE GENERAL PROBLEMS

Let us now consider problems involving the more general differential equation (1.1). We do not assume that the region is necessarily a square, or a rectangle, but we do assume that its boundary consists of horizontal and vertical lines such that for any mesh point in R the four neighboring points are in R or else lie on S. An example of such a region is shown in Figure 3.1.

We consider the difference equation obtained by the use of "symmetric" difference representations of the derivatives. Thus we let

$$(3.1)\qquad \frac{\partial}{\partial x}\left(A\,\frac{\partial u}{\partial x}\right) \sim \frac{1}{h}\left\{A\left(x+\frac{h}{2},y\right)\left[\frac{u(x+h,y)-u(x,y)}{h}\right] - A\left(x-\frac{h}{2},y\right)\left[\frac{u(x,y)-u(x-h,y)}{h}\right]\right\}$$

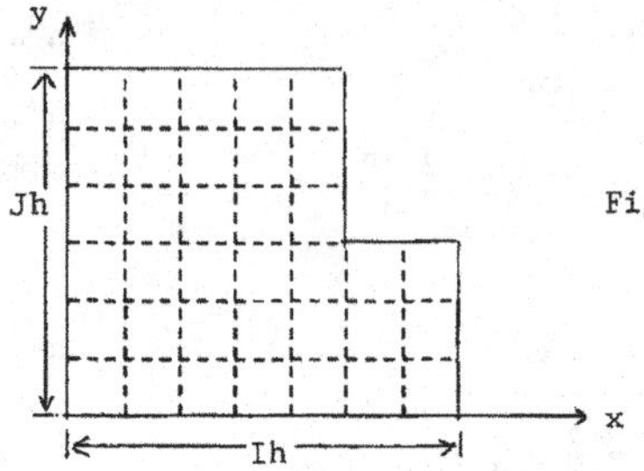

Figure 3.1

Substituting in (1.1) and solving for u(x,y) we get

$$(3.2)\qquad u(x,y) = \beta_1(x,y)u(x+h,y) + \beta_2(x,y)u(x,y+h) + \beta_3(x,y)u(x-h,y) + \beta_4(x,y)u(x,y-h) + \tau(x,y)$$

where

$$(3.3)\qquad \begin{cases} \beta_1(x,y) = \dfrac{A(x+\frac{h}{2},y)}{S(x,y)} & \beta_2(x,y) = \dfrac{C(x,y+\frac{h}{2})}{S(x,y)} \\ \beta_3(x,y) = \dfrac{A(x-\frac{h}{2},y)}{S(x,y)} & \beta_4(x,y) = \dfrac{C(x,y-\frac{h}{2})}{S(x,y)} \\ S(x,y) = A(x+\frac{h}{2},y) + A(x-\frac{h}{2},y) + C(x,y+\frac{h}{2}) + C(x,y-\frac{h}{2}) - h^2F(x,y) & \\ \tau(x,y) = -h^2G(x,y)/S(x,y) & \end{cases}$$

Young [17,19] gave the following bound for S(B).

$$(3.4)\qquad S(B) \le \frac{2(\bar{A}+\bar{C})}{2(\bar{A}+\bar{C})+h^2(\underline{-F})}\left\{1 - \frac{2\underline{A}\sin^2\frac{\pi}{2I} + 2\underline{C}\sin^2\frac{\pi}{2J}}{\frac{1}{2}(\bar{A}+\underline{A}) + \frac{1}{2}(\bar{C}+\underline{C}) + \frac{1}{2}(\bar{A}-\underline{A})\cos\frac{\pi}{I} + \frac{1}{2}(\bar{C}-\underline{C})\cos\frac{\pi}{J}}\right\}.$$

Here we let

$$(3.5)\qquad \underline{A} \le A(x,y) \le \bar{A}\,,\quad \underline{C} \le C(x,y) \le \bar{C},\quad (\underline{-F}) \le (-F)$$

If we let ω_b be determined by (2.21), then $RR(\mathcal{L}_{\omega_b}) = O(h^{-1})$.

4. THE SSOR METHOD

We now define the SSOR method for the system (2.5). Each iteration consists of two half iterations. In the first half iteration we get $u^{(n+\frac{1}{2})}$ using the forward SOR method. Thus

$$u^{(n+\frac{1}{2})} = \omega(Lu^{(n+\frac{1}{2})} + Uu^{(n)} + c) + (1-\omega)u^{(n)}. \tag{4.1}$$

In the second half iteration we get $u^{(n+1)}$ from $u^{(n+\frac{1}{2})}$ using the backward SOR method. We have

$$u^{(n+1)} = \omega(Lu^{(n+\frac{1}{2})} + Uu^{(n+1)} + c) + (1-\omega)u^{(n+\frac{1}{2})}. \tag{4.2}$$

Eliminating $u^{(n+\frac{1}{2})}$ we get

$$u^{(n+1)} = \mathcal{S}_\omega u^{(n)} + k_\omega, \tag{4.3}$$

where k_ω is a suitable vector and

$$\mathcal{S}_\omega = (I-\omega U)^{-1}(\omega L + (1-\omega)I)(I - \omega L)^{-1}(\omega U + (1-\omega)I). \tag{4.4}$$

Let us now assume that the matrix A of (2.5) is symmetric and positive definite and that we have bounds $\bar{\mu}$ and $\bar{\beta}$ for S(B) and S(LU), respectively, such that $\bar{\mu} \le 2\sqrt{\bar{\beta}}$. It can be shown (see, for instance, [20]) that, assuming also that A has Property A

$$S(\mathcal{S}_\omega) \le \begin{cases} 1 - \omega(2-\omega)\dfrac{1-\bar{\mu}}{1-\omega\bar{\mu}+\omega^2\bar{\beta}}, & \text{if } \bar{\beta} \ge \frac{1}{4} \text{ or } \bar{\beta} < \frac{1}{4} \text{ and } \omega < \omega^*. \\ 1 - \omega(2-\omega)\dfrac{1+\bar{\mu}}{1+\omega\bar{\mu}+\omega^2\bar{\beta}}, & \text{if } \bar{\beta} < \frac{1}{4} \text{ and } \omega > \omega^*. \end{cases} \tag{4.5}$$

Here for $\bar{\beta} < \frac{1}{4}$ we define ω^* by

$$\omega^* = \frac{2}{1+\sqrt{1-4\bar{\beta}}}\,. \tag{4.6}$$

The bounds (4.5) can be minimized by choosing ω by

$$\omega_1 = \begin{cases} \dfrac{2}{1+\sqrt{1-2\bar{\mu}+4\bar{\beta}}}, & \text{if } \bar{\mu} \le 4\bar{\beta} \\ \dfrac{2}{1+\sqrt{1-4\bar{\beta}}}, & \text{if } \bar{\mu} > 4\bar{\beta}\,. \end{cases} \tag{4.7}$$

The corresponding bounds on $S(\mathcal{S}_\omega)$ are

$$S(\mathcal{S}_{\omega_1}) \le \begin{cases} \dfrac{1 - \dfrac{1-\bar{\mu}}{\sqrt{1-2\bar{\mu}+4\bar{\beta}}}}{1 + \dfrac{1-\bar{\mu}}{\sqrt{1-2\bar{\mu}+4\bar{\beta}}}}, & \text{if } \bar{\mu} \le 4\bar{\beta} \\ \dfrac{1-\sqrt{1-4\bar{\beta}}}{1+\sqrt{1-4\bar{\beta}}} = \omega_1 - 1, & \text{if } \bar{\mu} > 4\bar{\beta}\,. \end{cases} \tag{4.8}$$

For the model problem it can be shown that $S(B) = \cos \pi h$, $S(LU) \le \frac{1}{4}\cos^2 \frac{\pi h}{2}$ so that

(4.9) $$\omega_1 = \frac{2}{1+\sqrt{3}\,\sin\frac{\pi h}{2}}$$

(4.10) $$S(\mathcal{S}_{\omega_1}) \le \frac{1-\frac{2}{\sqrt{3}}\sin\frac{\pi h}{2}}{1+\frac{2}{\sqrt{3}}\sin\frac{\pi h}{2}}.$$

Moreover, we have

(4.11) $$RR(\mathcal{S}_{\omega_1}) \doteq \frac{\sqrt{3}}{2\pi}\,h^{-1},$$

which is only slightly larger than $RR(\mathcal{L}_{\omega_b}) \doteq (2\pi h)^{-1}$.

The SSOR method would appear to converge slower than the SOR method even though it requires twice as much work per iteration. However, as we show in the next section, the use of an acceleration procedure (which cannot be used with the SOR method) results in much faster convergence for the SSOR method.

In order to obtain an order-of-magnitude increase in convergence rate it is necessary to show that ω can be chosen so that

(4.12) $$RR(\mathcal{S}_\omega) = O(h^{-1}).$$

This can be done for the difference equation (3.2) provided that the functions $A(x,y)$ and $C(x,y)$ belong to $C^{(2)}$. To determine ω_1 by (4.7) we let $\bar{\mu}$ be given by (3.4) and $\bar{\beta}$ by

(4.13) $$\bar{\beta} \le \max_{R_h} \{\beta_3(x,y)[\beta_1(x-h,y)+\beta_2(x-h,y)] + \beta_4(x,y)[\beta_1(x,y-h)+\beta_2(x,y-h)]\}.$$

Here R_h is the set of mesh points in R. It is shown in [20] that

(4.14) $$\bar{\beta} \le \frac{1}{4} + O(h^2).$$

Moreover, for the case $\bar{\beta} > \frac{1}{4}$ we can write, by (4.8),

(4.15) $$S(\mathcal{S}_{\omega_1}) \le \frac{1-\sqrt{\frac{1-\bar{\mu}}{2}}\left\{1+\frac{2(\bar{\beta}-1/4)}{1-\bar{\mu}}\right\}^{-1/2}}{1+\sqrt{\frac{1-\bar{\mu}}{2}}\left\{1+\frac{2(\bar{\beta}-1/4)}{1-\bar{\mu}}\right\}^{-1/2}}$$

Since $\bar{\mu} = 1 - c_1h^2 + O(h^4)$, by (3.4), it follows from (4.14) that

$$S(\mathcal{S}_{\omega_1}) \le 1 - kh + o(h)$$

for some positive constant k. Hence (4.12) holds.

5. CONVERGENCE ACCELERATION

If A is positive definite and if $0 < \omega < 2$, then the eigenvalues of the matrix $\mathcal{S}_\omega$ are real, nonnegative, and less than unity. Consequently, the convergence of the SSOR method (4.3) can be accelerated by an order-of-magnitude using variable extrapolation. We choose a positive integer m and the extrapolation factors $\theta_1,\theta_2,\ldots,\theta_m$ by

(5.1) $$\theta_k = \frac{1}{1-S(\mathcal{S}_\omega)\cos^2\frac{(2k-1)\pi}{4m}}, \qquad k = 1,2,\ldots,m.$$

We use the iterative method

(5.2) $$u^{(n+1)} = \theta_{n+1}(\mathcal{S}_\omega u^{(n)} + k) + (1-\theta_{n+1})u^{(n)}.$$

Here the θ_k are used in a cyclic order $\theta_1, \theta_2, \ldots, \theta_m, \theta_1, \theta_2, \ldots$. To determine the rapidity of convergence of the method we note that, by (5.2), we can write

(5.3) $$u^{(n)} = \mathcal{G}_n u^{(0)} + k_n$$

where $\mathcal{G}_n$ is a polynomial in $\mathcal{S}_\omega$. It can be shown that for any integer t,

(5.4) $$S(\mathcal{G}_{tm}) = \left(\frac{2r^{m/2}}{1+r^m}\right)^t$$

where

(5.5) $$r = \left(\frac{\sqrt{S(\mathcal{S}_\omega)}}{1+\sqrt{1-S(\mathcal{S}_\omega)}}\right)^4 .$$

The reciprocal average rate of convergence is

(5.6) $$RR_{tm}(\mathcal{G}_{tm}) = \left[-\frac{1}{m}\log\frac{2r^{m/2}}{1+r^m}\right]^{-1} .$$

The reciprocal asymptotic average rate of convergence is

(5.7) $$RR_\infty(\mathcal{G}_n) = \left[-\frac{1}{2}\log r\right]^{-1} \doteq \frac{1}{2}\sqrt{RR(\mathcal{S}_\omega)} ,$$

which is less than $RR(\mathcal{S}_\omega)$ by an order-of-magnitude. For the model problem using $\omega = \omega_1$ we have

(5.8) $$RR_\infty(\mathcal{G}_m) = \frac{3^{\frac{1}{4}}}{2\sqrt{2}\sqrt{\pi}} h^{-\frac{1}{2}} ,$$

which is much less than $RR(\mathcal{L}_{\omega_b}) \doteq (2\pi h)^{-1}$.

The choice of m is governed by several considerations. We require m to be large enough so that $RR_{tm}(\mathcal{G}_{tm})$ is at least some fraction, say 80%, of $RR_\infty(\mathcal{G}_n)$. On the other hand, we do not wish to make m too large, both because of possible numerical instability (see Young [16]) and also because one can only expect convergence after m, 2m, 3m, ... iterations.

As an alternative to the variable extrapolation procedure (5.2) we can use semi-iteration. (See Varga [14] and Golub and Varga [9].) Here at each step one determines $u^{(n+1)}$ from $u^{(n)}$ and $u^{(n-1)}$. The convergence of this procedure is faster than with variable extrapolation and there is no danger of instability. One disadvantage of the semi-iterative method is that an extra vector, namely $u^{(n-1)}$, is required to be retained in memory.

6. NUMERICAL RESULTS

In this section we describe some numerical experiments which were performed using the accelerated SSOR method. The differential equation

(6.1) $$\frac{\partial}{\partial x}\left(A\frac{\partial u}{\partial x}\right) + \frac{\partial}{\partial y}\left(C\frac{\partial u}{\partial y}\right) = 0$$

was solved in the unit square with boundary values zero on all sides except unity on the side $y = 0$. Various choices of $A(x,y)$ and $C(x,y)$ were used. (See Concus and Golub [3] and Axelsson [1].)

Accelerated SSOR methods, including variable extrapolation and semi-iteration, were used both with estimated optimum parameters and also with exact optimum parameters. For the procedure based on estimated parameters, we first estimated $S(B)$ and $S(LU)$ by (3.4) and (4.13), respectively. Then ω_1 and a bound for $S(\mathcal{S}_{\omega_1})$ were determined by (4.7) and (4.8), respectively. Next, for the variable extrapolation procedure m was determined so that

$$-\frac{1}{m}\log\frac{2r^{m/2}}{1+r^m} \geq (.8)\left(-\frac{1}{2}\log r\right).$$

Here r is given by (5.5). The extrapolation parameters were computed by (5.1). The iteration procedure (5.2) was carried out for t cycles of m iterations where

$$\left(\frac{2r^{m/2}}{1+r^m}\right)^t \leq \zeta = 10^{-6}.$$

When this condition is satisfied, it follows that, if $u^{(0)} = 0$,

$$\frac{\|u^{(tm)} - \bar{u}\|_{A^{\frac{1}{2}}}}{\|\bar{u}\|_{A^{\frac{1}{2}}}} \leq \zeta.$$

Here, for any vector v we let

$$\|v\|_{A^{\frac{1}{2}}} = \|A^{\frac{1}{2}}v\| = \sqrt{(v,Av)}\,.$$

A similar procedure was carried out using ω such that $S(\mathcal{S}_\omega)$ is minimized. The value of ω and the corresponding value of $S(\mathcal{S}_\omega)$ were found by determining $S(\mathcal{S}_\omega)$ for several values of ω by the power method. The SSOR semi-iterative (SSOR-SI) method was used as well as the extrapolated method with both sets of parameters. In addition, the SOR method was used with the exact value of the optimum ω. The number of iterations was determined by the formula

$$\mathcal{N} = \frac{\log \zeta^{-1}}{-\log(\omega_b - 1)}\,.$$

The actual number of iterations would normally be somewhat higher (see, for instance, Young [19, Chapter 7]).

Numerical results are given in Table 6.1. The following observations are made.

(1) With the accelerated SSOR methods the number of iterations varies as $h^{-\frac{1}{2}}$, even with the estimated parameters. With the SOR method, on the other hand, the number of iterations varies as h^{-1}. Even considering the fact that twice as much work per iteration is required for the SSOR method and in spite of the additional complication due to the acceleration process, there is a worthwhile saving using the SSOR method for problems involving small mesh sizes.

(2) A worthwhile increase in convergence rate can be obtained by using the optimum parameters. However, it would not be practical to expend too many extra iterations in search of the optimum parameters. This is discussed further in Section 7.

(3) In the cases shown, the use of semi-iteration rather than variable extrapolation results in a substantial saving in some cases. This is because with variable extrapolation one can only expect convergence in tm iterations, for some integer t. An alternative procedure is to choose a larger value of m such that convergence will occur in n iterations where n/m is an integer. Actually, one can let $m = n$, or if stability is a concern, one can let $m \sim m/s$ for some small integer s. On the other hand, if memory capacity is not a serious problem, the use of semi-iteration, rather than variable extrapolation, is recommended.

TABLE 6.1. NUMERICAL RESULTS

	h	Estimated SSOR-VE	Estimated SSOR-SI	Optimum SSOR-VE	Optimum SSOR-SI	Optimum SOR
$A = C = 1$	$\frac{1}{20}$	25	19	20	16	44
	$\frac{1}{40}$	35	26	30	23	88
	$\frac{1}{80}$	45	37	40	32	174
$A = C = e^{10(x+y)}$	$\frac{1}{20}$	12	10	12	10	24
	$\frac{1}{40}$	20	15	16	14	48
	$\frac{1}{80}$	25	21	25	20	119
$A = \frac{1}{1+2x^2+y^2}$	$\frac{1}{20}$	35	28	20	17	45
	$\frac{1}{40}$	50	40	30	23	90
$C = \frac{1}{1+x^2+2y^2}$	$\frac{1}{80}$	70	57	40	33	177
$A = C = \begin{cases} 1+x, & 0 \le x \le \frac{1}{2} \\ 2-x, & \frac{1}{2} \le x \le 1 \end{cases}$	$\frac{1}{20}$	24	21	20	17	46
	$\frac{1}{40}$	40	32	30	24	92
	$\frac{1}{80}$	60	49	40	33	180
$A = 1 + 4\lvert x - \frac{1}{2}\rvert^2$	$\frac{1}{20}$	35	28	25	19	43
$C = \begin{cases} 1, & 0 \le x < \frac{1}{2} \\ 9, & \frac{1}{2} \le x \le 1 \end{cases}$	$\frac{1}{40}$	50	40	30	25	86
	$\frac{1}{80}$	70	56	45	34	164
$A = 1 + \sin \frac{\pi(x+y)}{2}$	$\frac{1}{20}$	12	11	12	10	24
	$\frac{1}{40}$	20	15	20	15	47
$C = e^{10(x+y)}$	$\frac{1}{80}$	30	22	25	21	120

(Note that each SSOR iteration is approximately equivalent to two SOR iterations.)

7. DYNAMIC IMPROVEMENT OF PARAMETERS

As described in Section 6, the optimum ω and the corresponding value of $S(\mathcal{S}_\omega)$ can be found by determining $S(\mathcal{S}_\omega)$ for various values of ω using the power method. A more sophisticated scheme was used by Evans and Forrington [8] based on the analysis of Habetler and Wachspress [10]. This analysis yields the formulas

$$\omega = \frac{2}{1+\sqrt{1-2\alpha+4\beta}} \tag{7.1}$$

$$S(\mathcal{S}_\omega) = \left(1 - \frac{1-\alpha}{\sqrt{1-2\alpha+4\beta}}\right)\left(1 + \frac{1-\alpha}{\sqrt{1-2\alpha+4\beta}}\right)^{-1} \tag{7.2}$$

for the optimum value of ω and the corresponding $S(\mathcal{S}_\omega)$. Here v is a vector such that $(v,Dv) = 1$ and $\mathcal{S}_\omega v = S(\mathcal{S}_\omega)v$. Moreover, $\alpha = (v, \overset{\omega}{D}Bv)$, $\beta = (v, DLUv)$. Evans and Forrington used an iterative procedure based on (7.1) and (7.2) which worked quite well for Laplace's equation. However, our numerical results indicate that for certain cases where $S(LU) < 1/4$, the formulas (7.1) and (7.2) do not hold. In some cases, the heuristically derived formulas (which are motivated by (4.7)-(4.8)),

$$\omega = \frac{2}{1+\sqrt{1-4\beta}} \tag{7.3}$$

$$S(\mathcal{S}_\omega) = \frac{1-\sqrt{1-4\beta}}{1+\sqrt{1-4\beta}} = \omega - 1 \tag{7.4}$$

seem to hold. Further studies of these formulas are being made.

Unless a great many cases are to be solved with the same matrix, it is clear that it is not practical to devote very many iterations to finding the optimum parameters, unless such iterations are also "useful" in improving the approximate solution. Thus, for example, the determination of the optimum ω by the power method involves many "wasted" iterations and would not usually be practical. Efforts are now underway to obtain improved estimates of ω dynamically with a minimum number of extra iterations.

8. PREVIOUS WORK ON THE SSOR METHOD

The SSOR method and the variable extrapolation procedure described in Section 5 were introduced by Sheldon [12] in 1965. Sheldon conjectured that for Laplace's equation the convergence was $O(h^{-\frac{1}{2}})$. That such convergence could be attained with Laplace's equation was shown by Habetler and Wachspress [10]. However, they did not give an explicit procedure for finding satisfactory iteration parameters. Ehrlich [6,7] considered the line SSOR method and proved $O(h^{-\frac{1}{2}})$ convergence for Laplace's equation and the rectangle. Young [18,19] gave an explicit procedure for determining iteration parameters for both point and line SSOR leading to $O(h^{-\frac{1}{2}})$ convergence for Laplace's equation and for a class of regions not limited to rectangles. Axelsson [1] considered a generalization of the SSOR method where the relaxation factor ω varies from equation to equation. He showed $O(h^{-\frac{1}{2}})$ convergence.

9. OTHER METHODS

While $O(h^{-\frac{1}{2}})$ and even faster convergence has been obtained for some problems with other methods, the number of situations where such convergence can be rigorously proved to hold with reasonable generality is rather limited. For example, one can prove $O(\log h^{-1})$ convergence for the Peaceman-Rachford method [11] provided that the region is a rectangle and the equation is "separable" (see, for instance, Birkhoff, Varga, and Young [2]). However, in spite of many efforts the theory has not been extended to cover more general cases. An exception has been Widlund's treatment of nonseparable equations for the rectangle [15].

Dupont, Kendall, and Rachford [5] have shown $O(h^{-\frac{1}{2}})$ convergence for certain "almost factorization" procedures. (See also Stone [13] and Dupont [4].) We remark that the SSOR method can be considered as an "almost factorization" procedure.

10. ACKNOWLEDGEMENT

The author wishes to acknowledge the contribution of Vitalius Benokraitis of The University of Texas at Austin both in carrying out the numerical studies described above and in making useful suggestions concerning the theory. The cooperation of the University of Texas Computation Center in making its facilities available for the numerical work is also acknowledged.

BIBLIOGRAPHY

1. Axelsson, O. "A generalized SSOR method," BIT 13, 443-467 (1972).
2. Birkhoff, G., Varga, R. S., and Young, D. "Alternating direction implicit methods," Advances in Computers, Vol. 3, 189-273 (1962).
3. Concus, Paul, and Golub, Gene H. "Use of fast direct methods for the efficient numerical solution of nonseparable elliptic equations," Report STAN-CS-72-278, Computer Science Department, Stanford University, April 1972.
4. Dupont, Todd. "A factorization procedure for the solution of elliptic difference equations," SIAM J. Numer. Anal. 5, 753-782 (1968).
5. Dupont, Todd, Kendall, Richard P., and Rachford, H.H., Jr. "Approximate factorization procedure for solving self-adjoint elliptic difference equations," SIAM J. Numer. Anal. 5, 559-573 (1968).
6. Ehrlich, L. W. "The Block Symmetric Successive Overrelaxation Method," doctoral thesis, University of Texas, Austin (1963).
7. Ehrlich, L. W. "The block symmetric successive overrelaxation method," J. Soc. Indust. Appl. Math. 12, 807-826 (1964).
8. Evans, D. J., and Forrington, C. V. D. "An iterative process for optimizing symmetric overrelaxation," Comput. J. 6, 271-273 (1963).
9. Golub, G. H., and Varga, R. S. "Chebyshev semi-iterative methods, successive, overrelaxation iterative methods, and second-order Richardson iterative methods," Numer. Math., Parts I and II 3, 147-168 (1961).
10. Habetler, G. J., and Wachspress, E. L. "Symmetric successive overrelaxation in solving diffusion difference equations," Math. Comp. 15, 356-362 (1961).
11. Peaceman, D. W., and Rachford, H. H., Jr. "The numerical solution of parabolic and elliptic differential equations," J. Soc. Indus. Appl.Math. 3, 28-41 (1955).
12. Sheldon, J. "On the numerical solution of elliptic difference equations," Math. Tables Aids Comput. 9, 101-112 (1955).
13. Stone, H. L. "Iterative solution of implicit approximation of multidimensional partial differential equations," SIAM J. Numer. Anal. 5, 530-558 (1968).
14. Varga, R. S. "A comparison of the successive overrelaxation method and semi-iterative methods using Chebyshev polynomials," J. Soc. Indus. Appl. Math. 5, 39-46 (1957).
15. Widlund, O. B. "On the rate of convergence of an alternating direction implicit method in a noncommutative case," Math. Comp. 20, 500-515 (1966).
16. Young, D. M. "On the solution of linear systems by iteration," Proc. Sixth Symp. in Appl. Math. Amer. Math. Soc. VI, McGraw-Hill, New York, 283-298 (1956).
17. Young, D. M. "A bound on the optimum relaxation factor for the successive overrelaxation method," Numer. Math. 16, 408-413 (1971).
18. Young, D. M. "Second-degree iterative methods for the solution of large linear systems," J. Approx. Theory 5, 137-148 (1972).

19. Young, D. M. Iterative Solution of Large Linear Systems, Academic Press, New York, 1971.

20. Young, D. M. "On the accelerated SSOR method for solving large linear systems," to appear.

CONSTRAINED VARIATIONAL PRINCIPLES AND PENALTY FUNCTION METHODS IN FINITE ELEMENT ANALYSIS

O. C. ZIENKIEWICZ

ABSTRACT

Penalty functions are used to modify variational principles used in finite element analysis to enforce constraints. The procedure is found useful in imposing constraints implicit in the functional or to impose required interelement continuity. While the process is approximate, quite good practical results can be obtained if sufficient accuracy is available in the computer. The procedure is illustrated on several problems of interest in elasticity and fluid mechanics.

1. INTRODUCTION

'Standard' finite element discretisation of physical problems [1] has, on occasion, to be supplemented by imposing constraints on the variational principle governing the problem. This may be motivated by necessity - if, for instance, the validity of the principle is dependent on the constraint or simply convenience in avoiding stringent continuity requirements which may be imposed by integrability requirements on the (piecewise defined) trial functions.

The customary procedure of imposing such constraints by the use of Lagrange multipliers has been widely used but suffers the inconvenience of increasing the number of unknown parameters to be solved for, as well as, in linear problems, yielding indefinite matrices. An alternative method of imposing constraints by 'penalty functions' stems from literature on optimization [2] but appears not to have been used explicitly in the finite element process, although in the particular case of 'equality constraints' it appears particularly well suited to it avoiding both difficulties mentioned in connection with Lagrange multipliers.

Briefly if the variational principle in the unconstrained form requires the determination of a function (or set of functions) $\underset{\sim}{\varphi}$ which makes a functional

$$\chi = \chi(\varphi) \tag{1}$$

defined in a domain Ω and a boundary Γ, stationary, then if a set of constraints

$$\underset{\sim}{C}(\varphi) = 0 \qquad \text{on} \quad \Omega^c \text{ (or } \Gamma^c) \tag{2}$$

is introduced we modify the functional to

$$\bar{\chi} = \chi + \alpha \int_{\Omega^c/\Gamma^c} \underset{\sim}{C}(\underset{\sim}{\varphi})^2 \, d\Omega \tag{3}$$

and seek its stationary value. In equation (3) α is a large, positive, number - and clearly the larger it is, the closer is the satisfaction of constraint imposed. While the proposed method will, obviously, be best suited to situations for which a minimum of χ is sought, it is equally applicable for situations in which χ is merely stationary. The question of deciding what value of α is "sufficiently large" must in general be left to an empirical test. Obviously a purely numerical conflict may exist between the accuracy attainable and numerical conditioning of the discretised equations resulting, but fortunately with the large accuracy available in modern computing machines this is seldom serious.

As already mentioned, the main motivation for imposing constraints comes from two sources: first, their implicit requirement in the functional; second, the continuity requirement imposed on the trial functions. We shall discuss the application of the procedure under these two headings by suitably chosen examples. Some indication of yet unexplored possibilities will also be given.

2. CONSTRAINTS IMPLICIT IN THE FUNCTIONAL

2.1 Incompressible Elastic Behaviour - Displacement Formulation

Let $\underset{\sim}{\varepsilon}$ define the strain (infinitesimal) vector, $\underset{\sim}{\varepsilon}_o$ the initial strain and $\underset{\sim}{\sigma}$ be the elastic stress given by

$$\underset{\sim}{\sigma} = \underset{\sim}{D}(\underset{\sim}{\varepsilon} - \underset{\sim}{\varepsilon}_o) \tag{4}$$

where $\underset{\sim}{D}$ is a matrix of elastic constraints.

The problem of determining a displacement state with stresses which are in equilibrium with boundary tractions $\bar{t}$ and body forces $\underset{\sim}{b}$ is given as one of minimizing the potential energy with respect to variations of the displacement $\underset{\sim}{u}$

$$\chi = \frac{1}{2}\int_\Omega \underset{\sim}{\varepsilon}^T \underset{\sim}{D}\, \underset{\sim}{\varepsilon} \, d\Omega - \int_\Omega \underset{\sim}{\varepsilon}^T \underset{\sim}{D}\, \underset{\sim}{\varepsilon}_o \, d\Omega - \int_\Omega \underset{\sim}{u}^T \underset{\sim}{b} \, d\Omega - \int_{\Gamma_\sigma} \underset{\sim}{u}^T \bar{\underset{\sim}{t}} \, d\Gamma \tag{5}$$

with the strain related to displacement by a linear operator

$$\underset{\sim}{\varepsilon} = \underset{\sim}{L}\, \underset{\sim}{u} \, . \tag{6}$$

(Γ_σ is the boundary on which tractions are prescribed). With an incompressible material the matrix $\underset{\sim}{D}$ can not be determined and the formulation apparently fails. However, if we consider only the deviatoric components of stress $\underset{\sim}{\sigma}'$, we find that these can be uniquely determined as

$$\underset{\sim}{\sigma}' = \underset{\sim}{D}'(\underset{\sim}{\varepsilon} - \underset{\sim}{\varepsilon}_o) \tag{7}$$

in which $\underset{\sim}{D}'$ is determined.

As no work is done by the hydrostatic stress component during imposition of strains of constant volume, we can pose the problem as one of minimizing χ' in which $\underset{\sim}{D}$ is replaced by $\underset{\sim}{D}'$. However the displacements are now subject to the constraint

$$\text{div } \underset{\sim}{u} = 0 \ . \tag{8}$$

With the application of expression (3) we now minimize

$$\bar{\chi} = \chi' + \alpha \int_\Omega \text{div}^2 \ \underset{\sim}{u} \ d\Omega \ . \tag{9}$$

A little elaboration shows immediately that the large number α can be identified with the bulk modulus for an isotropic material case and that the procedure proposed is in fact identical with that often practiced by engineers by approaching incompressible formulation via conventional programs in which a very large bulk modulus (or Poisson's ratio → 0.5) was introduced. Such approaches have been shown to be generally just as effective [3] as those in which incompressibility is enforced by Lagrange multipliers or alternative variational forms [4], [5].

2.2 Incompressible Viscous Fluid Flow

Here it has become customary to introduce stream functions to ensure incompressibility in two dimensional situations - but unfortunately the process is not easily generalised for three dimensions (and even in two poses certain difficulties). The alternative of introducing Lagrange multipliers - identified with the 'pressure' is possible [1]. Here an alternative formulation parallelling that obtained for the elasticity problem will be followed.

If $\underset{\sim}{u}$ now represents the velocity vector and $\dot{\underset{\sim}{\varepsilon}}$ the rate of straining, the relation (6) with $\underset{\sim}{\varepsilon}$ replaced by $\dot{\underset{\sim}{\varepsilon}}$ remains valid. The constitutive law of viscous fluid flow once again determines only the deviatoric stress components as

$$\underset{\sim}{\sigma}' = \underset{\sim}{D} \ \dot{\underset{\sim}{\varepsilon}} \tag{10}$$

and with isotropic, incompressible fluids it is easy to show that we can write D as a diagonal matrix

$$\underset{\sim}{D} = \mu \begin{bmatrix} 2 & & & & & 0 \\ & 2 & & & & \\ & & 2 & & & \\ & & & 1 & & \\ & & & & 1 & \\ 0 & & & & & 1 \end{bmatrix} \tag{11}$$

where μ is the viscosity coefficient.

If inertia effects are ignored for the moment (these can always be incorporated at the final stage of approximation via the body force term), we can consider the rate of virtual work done and arrive at a functional similar in all respects to that given by (5) but in which $\dot{\underset{\sim}{\varepsilon}}_o$ terms are omitted. Immediately we can write the complete formulation as that of finding the minimum of the functional

$$\bar{\chi} = \frac{1}{2}\int_\Omega \dot{\underset{\sim}{\varepsilon}}^T \mu \underset{\sim}{\varepsilon}\, d\Omega - \int_\Omega \underset{\sim}{u}^T \underset{\sim}{b}\, d\Omega - \int_{\Gamma_\sigma} \underset{\sim}{u}^T \bar{\underset{\sim}{t}}\, d\Gamma + \alpha \int_\Omega \mathrm{div}^2 \underset{\sim}{u}\, d\Omega \tag{12}$$

with

$$\dot{\underset{\sim}{\varepsilon}} = \underset{\sim}{L}\, \underset{\sim}{u} \,. \tag{13}$$

The process is thus formally identical to that of solving the previous solid mechanics problem (indeed simple replacement of the shear modulus G by the viscosity μ allows any two or three dimensional program developed for incompressible elasticity to be immediately used for the solution of creeping flow viscous problems).

We could here simply state the general preposition that follows from this, which is simply that: <u>the stress and velocity distribution in creeping, viscous flow is identical to the stress and displacement distribution in an incompressible elastic situation for which all prescribed displacement boundary conditions have been replaced by the prescribed velocity distributions on these boundaries</u>.*

2.3 <u>Three Dimensional - Magneto-Static Problem</u>

Here we start from the basic governing equations of the problem which can be written as

$$\mathrm{curl}\, \underset{\sim}{H} = \underset{\sim}{J} \qquad \text{a)}$$

$$\underset{\sim}{B} = \mu\, \underset{\sim}{H} \qquad \text{b)} \quad (14)$$

$$\mathrm{div}\, \underset{\sim}{B} = 0 \qquad \text{c)}$$

in which $\underset{\sim}{B}$ is the magnetic induction, $\underset{\sim}{H}$ magnetic field strength, μ - magnetic permeability and $\underset{\sim}{J}$ the current density.

Defining a vector potential $\underset{\sim}{A}$ such that

$$\underset{\sim}{B} = \mathrm{curl}\, \underset{\sim}{A} \tag{15}$$

we satisfy identically equation (14c) and on elimination of $\underset{\sim}{H}$ obtain the governing equation

$$\mathrm{curl}\, \frac{1}{\mu}\, \mathrm{curl}\, \underset{\sim}{A} = \underset{\sim}{J} \tag{16}$$

which, together with appropriate boundary conditions should define the problem. The operator is self adjoint and we can write the variational functional as

$$\chi = \int_\Omega \left(\frac{1}{2} \underset{\sim}{A}^T\, \mathrm{curl}\, \frac{1}{\mu}\, \mathrm{curl}\, \underset{\sim}{A} - \underset{\sim}{A}^T \underset{\sim}{J}\right) d\Omega \,. \tag{17}$$

A difficulty which arises immediately, however, is associated with an apparent non-uniqueness of $\underset{\sim}{A}$ as

---

*In both problems 2.1 and 2.2 by definition stresses are indeterminate within a hydrostatic component. However, boundary tractions will be obtained via "reactions" as in standard solid mechanics situation. It is this indeterminacy of stress that accounts for generally poor results obtained for stresses with high Poisson's ratio.

$$\text{curl}\ (\underset{\sim}{A}) \equiv \text{curl}\ (\underset{\sim}{A} + \text{grad}\ \varphi)$$

where φ is an arbitrary function. It is customary therefore to impose an additional constraint and require that

$$\text{div}\ \underset{\sim}{A} = 0\ . \tag{18}$$

Once again we can state the problem by the use of a modified functional

$$\bar{\chi} = \chi + \alpha \int_\Omega \text{div}^2\ \underset{\sim}{A}\ d\Omega \tag{19}$$

for discretisation. Numerical experience with this problem [6] has shown that, surprisingly, it is possible to obtain nonsingular results with a finite element discretisation without the constraint but that results are considerably improved by its introduction.

The last example in this section concerns a somewhat different problem of solid mechanics.

2.4 Complementary Energy Formulation with Equilibrium Imposed as a Constraint

It is well known that for stress fields which are in equilibrium we can obtain a solution satisfying strain-displacement compatibility by minimizing the complementary potential energy.

This can be defined, for an elastic solid, using the symbols of Section 2.1 as

$$\chi = \frac{1}{2} \int_\Omega \underset{\sim}{\sigma}^T \underset{\sim}{D}^{-1} \underset{\sim}{\sigma}\ d\Omega - \int_{\Gamma_u} \underset{\sim}{t}^T \bar{\underset{\sim}{u}}\ d\Gamma \tag{20}$$

where Γ_u is the part of the boundary on which displacements are prescribed.

While $\underset{\sim}{\sigma}$ and $\underset{\sim}{t}$ are related by simple geometric requirements, the practical application of this principle is limited by the difficulty of specifying equilibrating stress fields in a piecewise manner. Use of stress functions makes this possible under certain circumstances [7], [8], but even here difficulties remain.

In the present context we shall overcome the difficulty by imposing the equilibrium equations as a constraint. These equations can be written as

$$\underset{\sim}{L}^T \underset{\sim}{\sigma} + \underset{\sim}{b} = 0 \quad \text{in}\ \Omega\ , \qquad \underset{\sim}{G}^T \underset{\sim}{\sigma} - \bar{\underset{\sim}{t}} = 0 \quad \text{on}\ \Gamma_\sigma \tag{21}$$

where $\underset{\sim}{L}^T$ and $\underset{\sim}{G}^T$ are again appropriate linear operators ($\underset{\sim}{L}$ is identical with the operator defining strains as in eq. (5)). Immediately we can write as the new constrained functional

$$\bar{\chi} = \chi + \alpha' \int_\Omega (\underset{\sim}{L}^T \underset{\sim}{\sigma} + \underset{\sim}{b})^T (\underset{\sim}{L}^T \underset{\sim}{\sigma} + \underset{\sim}{b})\ d\Omega + \alpha \int_{\Gamma_\sigma} (\underset{\sim}{G}^T \underset{\sim}{\sigma} - \bar{\underset{\sim}{t}})^T (\underset{\sim}{G}^T \underset{\sim}{\sigma} - \bar{\underset{\sim}{t}})\ d\Gamma\ . \tag{22}$$

The possibilities offered by this new formulation appear excellent as the solution now concentrates on the unknowns of primary engineering interest, i.e., the stresses.

Further, very simple continuity requirements are now needed and are decided by the derivatives contained in $\underset{\sim}{L}$.

If for instance we consider a plane stress problem and for simplicity assume that the nodal parameter values are identified with the stresses

$$\underset{\sim}{\sigma}^T = [\sigma_{xx}, \sigma_{yy}, \sigma_{xy}] \tag{23}$$

then taking the trial expansion

$$\underset{\sim}{\sigma} = \sum \underset{\sim}{N}_i \underset{\sim}{\sigma}_i = \underset{\sim}{N} \underset{\sim}{\hat{\sigma}} \tag{24}$$

we have to minimize approximately the functional

$$\bar{\chi} = \frac{1}{2} \underset{\sim}{\hat{\sigma}}^T \left(\int_\Omega \underset{\sim}{N}^T \underset{\sim}{D}^{-1} \underset{\sim}{N} \, dxdy \right) \underset{\sim}{\hat{\sigma}} + \alpha \int_\Omega \left[(\underset{\sim}{L}^T \underset{\sim}{N}) \underset{\sim}{\hat{\sigma}} + \underset{\sim}{b}) \right]^T (\underset{\sim}{L}^T \underset{\sim}{N} \underset{\sim}{\hat{\sigma}} + \underset{\sim}{b}) \, dxdy \tag{25}$$

with

$$\underset{\sim}{L}^T = \begin{bmatrix} \frac{\partial}{\partial x}, & 0, & \frac{\partial}{\partial y} \\ 0, & \frac{\partial}{\partial y}, & \frac{\partial}{\partial x} \end{bmatrix} \tag{26}$$

assuming that the parameters $\underset{\sim}{\hat{\sigma}}$ were so chosen as to satisfy prescribed tractions on the boundaries. This leads to a set of linear equations

$$\underset{\sim}{K} \underset{\sim}{\hat{\sigma}} + \underset{\sim}{f} = 0 \tag{27}$$

with

$$\underset{\sim}{K}_{ji} = \sum \left[\int_{\Omega^e} \underset{\sim}{N}_j^T \underset{\sim}{D}^{-1} \underset{\sim}{N}_i \, dxdy + \alpha \int_{\Omega^e} (\underset{\sim}{L} \underset{\sim}{N}_j^T)(\underset{\sim}{L}^T \underset{\sim}{N}_i) \, dxdy \right] \tag{28}$$

in which the integration is, as usual, carried out element by element and the summation extends over all elements.

Practical application of above formulation is new and is being explored at the time of writing for problems of two and three dimensional stress analysis as well as for plates and shells.

3. CONSTRAINTS IMPOSED TO SATISFY CONTINUITY REQUIREMENTS

While the imposition of C^0 continuity presents little difficulties in the finite element process, C^1 continuity is difficult to achieve in piecewise defined fields of two dimensions (e.g. plate and shell problems) and impracticable in three dimensional situations. This has led to "variational crimes" reported by Strang [9] in which continuity requirements were sometimes relaxed. Despite the fact that here often 'crime pays', sometimes errors are introduced by this unjustified simplification and imposition of constraints to restore inter-element continuity has improved results. Work of Harvey and Kelsey [10], for instance, should be quoted in this context in which Lagrange multipliers are used with success to impose constraints on the "undeserving", incompatible, triangle introduced by Bazeley et al. in 1965 [11].

It appears possible to use the process of penalty functions to enforce continuity constraints in similar cases.

Consider for instance any plate or shell problems in which a discontinuous (non-compatible) displacement assumption was made on the generalized displacements. Let the inter-element continuity requirement be expressed as

$$\underset{\sim}{C}_L \underset{\sim}{u}_L + \underset{\sim}{C}_R \underset{\sim}{u}_R = 0 \tag{29}$$

where L and R refer to elements on either side of an interface.

The potential energy function χ has to be now supplemented by the requirement (29) in the form

$$\bar{\chi} = \chi + \alpha \int_I (\underset{\sim}{C}_L \underset{\sim}{u}_L + \underset{\sim}{C}_R \underset{\sim}{u}_R)^T (\underset{\sim}{C}_L \underset{\sim}{u}_L + \underset{\sim}{C}_R \underset{\sim}{u}_R)\, dI \tag{30}$$

where I is the interface between all elements. It will readily be seen that terms arising from minimization of the second term will simply supplement the element stiffness matrices obtained by incompatible assumptions.

In order to associate integrands I with elements (so as to maintain usual additive element properties) it is convenient to introduce the constraint in a slightly different manner.

Consider for instance the incompatible triangle of Ref. [11] in which due to cubic polynomial expansion used and due to the prescription of slopes and displacements of nodes, the only continuity violated is that of interelement slope. As this slope varies parabolically along an interface if we force it to assume the value given by the average of nodal slope at the centre of each side, then overall continuity will be achieved. We could then require simply that for each <u>element</u>

$$\underset{\sim}{A}\, \underset{\sim}{\hat{u}}^c - \underset{\sim}{B}\, \underset{\sim}{\hat{u}}^c = 0$$

where $\underset{\sim}{\hat{u}}^c$ are nodal displacement vectors for the element and $\underset{\sim}{A}$ and $\underset{\sim}{B}$ are matrices defining respectively the average values of appropriate nodal slope parameter and the mid side slope respectively. The modification of element stiffnesses would now be obtained by minimizing the following term and would not require an integration:

$$\alpha \quad (\underset{\sim}{A}\, \underset{\sim}{\hat{u}}^c - \underset{\sim}{B}\, \underset{\sim}{\hat{u}}^c)^T (\underset{\sim}{A}\, \underset{\sim}{\hat{u}}^c - \underset{\sim}{B}\, \underset{\sim}{u}^c)\ . \tag{31}$$

The success (or otherwise) of this possibility is being now investigated.

4. CONCLUDING REMARKS

The penalty function approach appears to be a viable and useful method of imposing constraints in the finite element context. Its use in many cases has unwittingly been already made. For instance the simple process of imposing prescribed boundary values by addition of a "large number" to diagonal matrix terms which is practised in many computer systems for finite elements [1] is simply one of its manifestations [12]. Some new applications have been here illustrated but further possibilities are quite large. Amongst these we shall mention such problems as plastic flow and creep deformation requiring incompressibility and 'inextensional' shell formulation, but the reader will recognize others.

ACKNOWLEDGEMENT

The author would like to express his thanks to Northwestern University for providing him with the necessary background during his sabbatical to accomplish this and other work.

REFERENCES

1. O. C. Zienkiewicz, The Finite Element Method in Engineering Science, McGraw-Hill, London, 1971.

2. R. H. Gallagher and O. C. Zienkiewicz, editors, Optimum Structural Design, J. Wiley & Sons, London, 1973.

3. G. Treharne, Ph.D. thesis, University of Wales Swansea, 1972.

4. L. R. Herrmann, "Elasticity Equations for Incompressible or Nearly Incompressible Materials by a Variational Theorem," AIAA Journal 3, 1965, p.1896-1900.

5. R. L. Taylor, K. S. Pister, and L. R. Herrmann, "On an Variational Theorem for Incompressible and Nearly Incompressible Orthotropic Elasticity," Int. J. Solids Structures 4, 1968, pp. 875-883.

6. J. Diserens and W. Trowbridge, Rutherford High Energy Laboratory, Private Communication, 1970.

7. B. Fraeijs de Veubeke and O. C. Zienkiewicz, "Strain Energy Bounds in Finite Element Analysis by Slab Analogy," J. Strain Analysis 2, 1967, pp. 265-271.

8. G. Sander, "Application of the Dual Analysis Principle," Proc. IUTAM Symposium on High Speed Computing of Elastic Structures, University of Liège, 1970, pp. 167-209.

9. G. Strang, "Variational Crimes in the Finite Element Method," from The Mathematical Foundations of the Finite Element Method with Application to Partial Differential Equations," editor A. K. Aziz, Academic Press, 1972, p. 689.

10. J. W. Harvey and S. Kelsey, "Triangular Plate Bending Elements with Enforced Compatibility," AIAA Journal 9, 1971, pp. 1023-1026.

11. G. P. Bazeley, Y. K. Cheung, B. M. Irons, and O. C. Zienkiewicz, "Triangular Elements in Plate Bending - Conforming and Nonconforming Solutions," Proc. Conf. on Matrix Methods in Structural Mechanics, Wright-Patterson Air Force Base, Ohio, 1965.

12. J. Campbell, University of Wales Swansea, Private Communication, 1972.

FINITE ELEMENT METHODS FOR PARABOLIC EQUATIONS

MILOŠ ZLÁMAL

1. Introduction

In the lecture we are speaking about results which we have obtained just recently. Exact statements of these results and proofs will be published elsewhere.

The finite element method was applied by the engineers for the solution of heat conduction problems a number of years ago. Their idea is that in the space dimension a finite element discretization is used whereas in time a finite difference method is applied. We analyze two methods: one is well known and makes use of the Crank - Nicolson discretization in time, the other is new and uses the Calahan discretization.

First we consider the linear initial-boundary value problem

$$\frac{\partial u}{\partial t} = Lu \qquad \text{in } \Omega \times (0,T),$$

$$(1) \qquad u = 0 \qquad \text{on } \Gamma \times \langle 0,T\rangle,$$

$$u|_{t=0} = g(x) \qquad \text{in } \Omega .$$

Here

$$Lu = \sum_{i,j=1}^{N} \frac{\partial}{\partial x_i}\left(a_{ij}(x)\frac{\partial u}{\partial x_j}\right) - a(x)u$$

and $x = (x_1,\dots,x_N)$ is a point of a bounded domain Ω in Euclidean N-space R^N with a smooth boundary Γ. As usual we require that

$$a_{ij}(x) = a_{ji}(x), \quad \sum_{i,j=1}^{N} a_{ij}(x)\xi_i\xi_j \geq \alpha \sum_{i=1}^{N}\xi_i^2, \quad \alpha = const > 0, \quad a(x) \geq 0.$$

At this place let us introduce some notations. The norm

$\|\cdot\|_{H^m}$ of the Sobolev space $H^m = W_2^{(m)}(\Omega)$ ($m = 0,1 \ldots$) and the inner product are denoted by $\|\cdot\|_m$ and $(.,.)_m$, respectively. H_0^1 is the closure of $\mathcal{D}(\Omega)$, the set of infinitely differentiable functions with compact support in Ω, in the $\|\cdot\|_1$ -norm.

The finite element discretization is considered in spaces V_h^p ($p = 1,2,\ldots$) depending on a small positive parameter h which are finite-dimensional subspaces of H_0^1 and which possess the following approximation property: to any $u \in H^{p+1} \cap H_0^1$ there exists a function $\hat{u} \in V_h^p$ such that

(2) $$\| u - \hat{u}\|_j \le Ch^{p+1-j}\|u\|_{p+1}, \quad j = 0,1.$$

Instead of (1) we solve the variational problem to find a function $u \in H_0^1$ such that besides the initial condition it satisfies for $t > 0$

$$(\dot{u}, \varphi)_0 + a(u,\varphi) = 0 \qquad \forall \varphi \in H_0^1.$$

Here $a(u,\varphi)$ is the energy bilinear functional

$$a(u,\varphi) = \int_\Omega \Big[\sum_{i,j=1}^{N} a_{ij}(x) \frac{\partial u}{\partial x_i}\frac{\partial \varphi}{\partial x_j} + a(x) u \varphi\Big] dx.$$

The continuous time Galerkin approximation of the solution u is a function $U(x,t) \in V_h^p$ such that

(3) $$\begin{aligned} (\dot{U},\varphi)_0 + a(U,\varphi) &= 0 \qquad \forall \varphi \in V_h^p \\ U(x,0) &= \hat{g}(x). \end{aligned}$$

The function $\hat{g}(x)$ is chosen according to (2). If we set $U(x,t) = \sum_{i=1}^{\nu} \alpha_i(t) v_i(x)$ where $\{v_i(x)\}_{i=1}^{\nu}$ form a basis of the space V_h^p we easily see that (3) is equivalent to the system of ordinary differential equations

(4) $$\dot{\boldsymbol{\alpha}} = -A\boldsymbol{\alpha}, \quad \boldsymbol{\alpha}(0) = \hat{\boldsymbol{g}}.$$

Here $\boldsymbol{\alpha}(t)$ is the vector $(\alpha_1(t),\ldots,\alpha_\nu(t))^T$, $A = M^{-1}K$, M is the mass matrix $\{(v_i,v_j)_0\}_{i,j=1}^{\nu}$, K is the stiffness matrix $\{a(v_i,v_j)\}_{i,j=1}^{\nu}$

and $\hat{g}$ is the vector such that $\hat{g}(x) = \sum_{i=1}^{\nu} \hat{g}_i v_i(x)$.

In general, the only way to compute $U(x,t)$ consists in the numerical solution of (4). We get approximate values of $U(x,t)$ for time levels $t = nk$, $n = 1,2,\ldots, n \leq Tk^{-1}$, k being the time increment. We denote these values by U^n and we take them as the approximate values of the exact values $u(x,nk) \equiv u^n$.

2. Crank - Nicolson discretization

The Crank-Nicolson discretization is equivalent to the solution of the system (4) by the trapezoidal method. Application of this method gives the recurrence relation

$$(5) \qquad (M + \tfrac{1}{2} kK)\alpha^{n+1} = (M - \tfrac{1}{2} kK)\alpha^n , \quad \alpha^0 = \hat{g} ;$$

here α^n is the approximation of the vector $\alpha(nk)$. The equivalent variational formulation of (5), called by Douglas and Dupont (1970) Crank-Nicolson-Galerkin approximation, has the form

$$(U^{n+1} - U^n, \varphi)_0 + \tfrac{1}{2} ka(U^{n+1} + U^n, \varphi) = 0 \qquad \forall \varphi \in V_h^p ,$$
$$U^0 = \hat{g}(x).$$

Our error bound, valid without any restriction on the size of h and k, is in the energy norm:

$$(6) \qquad \max_n \|u^n - U^n\|_1 \leq C(h^p + k^2)\|g\|_m , \quad m = \max(p+3,6).$$

The estimates of Douglas and Dupont, which were derived for a nonlinear equation introduced in the last section, are in a different norm whereas the order, both in h and k, is the same. Bramble and Thomée (1973) consider Galerkin methods with parameters h and k tied together by the relation kh^{-2} = const. Their Theorem 2, when applied to the Crank-Nicolson discretization, gives error bounds of the same order in k, again in a different norm.

3. Calahan discretization

The eigenvalues of the matrix -A are negative, however not bounded as the parameter h goes to zero. Therefore (4) belongs to stiff systems (see, e.g., Gear (1971)). For such systems we cannot use any numerical method unless the size of the increment k is restricted considerably. However, the A-stable methods (see Gear(1971)) do not require any restriction of the size of k. It is known that the trapezoidal method is a A-stable method. In fact, Dahlquist (1963) proved that a multistep method that is A-stable cannot have order greater than two and that among second-order A-stable methods the trapezoidal method is the best. Nevertheless we know A-stable methods of an arbitrary order and it is advisable to apply just these methods when we want to solve (4) with a higher order accuracy.

A third order A-stable method is the Calahan method (see Gear (1971), p.223). For the system (4) it gives

$$(7)\qquad \begin{aligned} &(M + bkK)\boldsymbol{\beta}^{n+1} = -kK\boldsymbol{\alpha}^{n} \\ &(M + bkK)\boldsymbol{\gamma}^{n+1} = -kK\boldsymbol{\alpha}^{n} + \beta kK\boldsymbol{\beta}^{n+1} \\ &\boldsymbol{\alpha}^{n+1} = \boldsymbol{\alpha}^{n} + \tfrac{1}{4}(3\boldsymbol{\beta}^{n+1} + \boldsymbol{\gamma}^{n+1}), \quad b = \tfrac{1}{2}(1 + \tfrac{1}{3}\sqrt{3}), \\ &\beta = \tfrac{2}{3}\sqrt{3}. \end{aligned}$$

The corresponding variational formulation is

$$(8)\qquad \begin{aligned} &(W^{n+1},\varphi)_0 + bka(W^{n+1},\varphi) = -ka(U^n,\varphi), \\ &(Z^{n+1},\varphi)_0 + bka(Z^{n+1},\varphi) = -ka(U^n,\varphi) + \beta ka(W^{n+1},\varphi), \\ &U^{n+1} = U^n + \tfrac{1}{4}(3W^{n+1} + Z^{n+1}). \end{aligned}$$

We have proved that the method is of the third order with respect to k:

$$\max_n \| u^n - U^n \| \leq C(h^p + k^3)\| g\|_m , \qquad m = \max(p+1,8); \tag{9}$$

here

$$\|\cdot\|^2 = \|\cdot\|_0^2 + k\,\|\cdot\|_1^2 .$$

Let us compare the amount of the main arithmetic operations necessary to carry out the procedures (5) and (7). Setting $\alpha^{n+1} = 2\vartheta^{n+1} - \alpha^n$ we can write (5) in the form

$$(M + \tfrac{1}{2}kK)\vartheta^{n+1} = M\alpha^n, \quad \alpha^{n+1} = 2\vartheta^{n+1} - \alpha^n .$$

Hence the main arithmetic operations necessary to carry out the first method consist of two parts: 1) We have to compute the matrices M and K and to carry out the forward elimination for the matrix $M+\frac{1}{2}kK$. 2) At every time step we have to multiply the matrix M by a vector and to carry out the back substitution. For the other method the first part is the same (due to the fact that we have the same matrix M+bkK at the left-hand side of the first two equations in (7)) whereas the second part is only twice as large. This is certainly a favourable result and we can expect that the procedure (7) will prove itself useful in applications.

4. A nonlinear problem

Douglas and Dupont (1970) derived error bounds for the case that the operator Lu is mildly nonlinear:

$$Lu = \sum_{i,j=1}^{N} \frac{\partial}{\partial x_i}\left(a_{ij}(x,u)\frac{\partial u}{\partial x_j}\right).$$

They assume that the coefficients $a_{ij}(x,u)$ satisfy a Lipschitz condition with respect to u, i.e. that for all $x \in \Omega$, r, $s \in R^1$ and i, j = 1, ..., N

$$|a_{ij}(x,r) - a_{ij}(x,s)| \leq L|r-s| ,$$

and that the functional

$$a(w;\, u,\varphi) = \int_{\Omega} \sum_{i,j=1}^{N} a_{ij}(x,w) \frac{\partial u}{\partial x_i} \frac{\partial \varphi}{\partial x_j}\, dx$$

is uniformly bounded from below and above, i.e. for all $x \in \Omega$, $\xi \in R^N$ and $s \in R^1$

$$\alpha \sum_{i=1}^{N} \xi_i^2 \le \sum_{i,j=1}^{N} a_{ij}(x,s)\, \xi_i \xi_j \le \alpha^{-1} \sum_{i=1}^{N} \xi_i^2 , \quad \alpha = const > 0.$$

They define the Crank-Nicolson-Galerkin approximation by the equation

$$(U^{n+1} - U^n, \varphi)_0 + ka\left(\frac{U^{n+1}+U^n}{2} ;\ \frac{U^{n+1}+U^n}{2}, \varphi \right) = 0 \qquad \forall \varphi \in V_h^p$$

which does not follow from the application of the trapezoidal method. They also linearize this procedure. One of the linearized procedures, called the Crank-Nicolson extrapolation, is defined as follows:

$$(U^{n+1} - U^n, \varphi)_0 + ka\left(\tfrac{3}{2}U^n - \tfrac{1}{2}U^{n-1};\ \frac{U^{n+1}+U^n}{2}, \varphi \right) = 0 \qquad \forall \varphi \in V_h^p.$$

U^1 requires, of course, a different definition. We define U^1 by

$$(U^1 - U^0, \varphi)_0 + ka\left(U^0;\ \frac{U^1+U^0}{2}, \varphi\right) = 0 \qquad \forall \varphi \in V_h^p.$$

We have proved that in both cases

$$\max_n \| u^n - U^n \|_0 \le C(h^{p+1} + k^2)$$

(now the constant C depends on the exact solution u).

References

Bramble, J.H. and Thomée, V., Discrete time Galerkin methods for a parabolic boundary-value problem. Annali di Mat. Pura Appl. (to appear).

Dahlquist, G., A special stability problem for linear multistep methods, BIT 3, 2743 (1963).

Douglas,J. Jr. and Dupont, T., Galerkin methods for parabolic equations, SIAM J. on Num.Anal. 7, 575-626 (1970).

Gear, C.W., Numerical initial value problems in ordinary differential equations, Prentice - Hall (1971).

Vol. 215: P. Antonelli, D. Burghelea and P. J. Kahn, The Concordance-Homotopy Groups of Geometric Automorphism Groups. X, 140 pages. 1971. DM 16,-

Vol. 216: H. Maaß, Siegel's Modular Forms and Dirichlet Series. VII, 328 pages. 1971. DM 20,-

Vol. 217: T. J. Jech, Lectures in Set Theory with Particular Emphasis on the Method of Forcing. V, 137 pages. 1971. DM 16,-

Vol. 218: C. P. Schnorr, Zufälligkeit und Wahrscheinlichkeit. IV, 212 Seiten. 1971. DM 20,-

Vol. 219: N. L. Alling and N. Greenleaf, Foundations of the Theory of Klein Surfaces. IX, 117 pages. 1971. DM 16,-

Vol. 220: W. A. Coppel, Disconjugacy. V, 148 pages. 1971. DM 16,-

Vol. 221: P. Gabriel und F. Ulmer, Lokal präsentierbare Kategorien. V, 200 Seiten. 1971. DM 18,-

Vol. 222: C. Meghea, Compactification des Espaces Harmoniques. III, 108 pages. 1971. DM 16,-

Vol. 223: U. Felgner, Models of ZF-Set Theory. VI, 173 pages. 1971. DM 16,-

Vol. 224: Revêtements Etales et Groupe Fondamental. (SGA 1). Dirigé par A. Grothendieck XXII, 447 pages. 1971. DM 30,-

Vol. 225: Théorie des Intersections et Théorème de Riemann-Roch. (SGA 6). Dirigé par P. Berthelot, A. Grothendieck et L. Illusie. XII, 700 pages. 1971. DM 40,-

Vol. 226: Seminar on Potential Theory, II. Edited by H. Bauer. IV, 170 pages. 1971. DM 18,-

Vol. 227: H. L. Montgomery, Topics in Multiplicative Number Theory. IX, 178 pages. 1971. DM 18,-

Vol. 228: Conference on Applications of Numerical Analysis. Edited by J. Ll. Morris. X, 358 pages. 1971. DM 26,-

Vol. 229: J. Väisälä, Lectures on n-Dimensional Quasiconformal Mappings. XIV, 144 pages. 1971. DM 16,-

Vol. 230: L. Waelbroeck, Topological Vector Spaces and Algebras. VII, 158 pages. 1971. DM 16,-

Vol. 231: H. Reiter, L^1-Algebras and Segal Algebras. XI, 113 pages. 1971. DM 16,-

Vol. 232: T. H. Ganelius, Tauberian Remainder Theorems. VI, 75 pages. 1971. DM 16,-

Vol. 233: C. P. Tsokos and W. J. Padgett. Random Integral Equations with Applications to stochastic Systems. VII, 174 pages. 1971. DM 18,-

Vol. 234: A. Andreotti and W. Stoll. Analytic and Algebraic Dependence of Meromorphic Functions. III, 390 pages. 1971. DM 26,-

Vol. 235: Global Differentiable Dynamics. Edited by O. Hájek, A. J. Lohwater, and R. McCann. X, 140 pages. 1971. DM 16,-

Vol. 236: M. Barr, P. A. Grillet, and D. H. van Osdol. Exact Categories and Categories of Sheaves. VII, 239 pages. 1971. DM 20,-

Vol. 237: B. Stenström, Rings and Modules of Quotients. VII, 136 pages. 1971. DM 16,-

Vol. 238: Der kanonische Modul eines Cohen-Macaulay-Rings. Herausgegeben von Jürgen Herzog und Ernst Kunz. VI, 103 Seiten. 1971. DM 16,-

Vol. 239: L. Illusie, Complexe Cotangent et Déformations I. XV, 355 pages. 1971. DM 26,-

Vol. 240: A. Kerber, Representations of Permutation Groups I. VII, 192 pages. 1971. DM 18,-

Vol. 241: S. Kaneyuki, Homogeneous Bounded Domains and Siegel Domains. V, 89 pages. 1971. DM 16,-

Vol. 242: R. R. Coifman et G. Weiss, Analyse Harmonique Non-Commutative sur Certains Espaces. V, 160 pages. 1971. DM 16,-

Vol. 243: Japan-United States Seminar on Ordinary Differential and Functional Equations. Edited by M. Urabe. VIII, 332 pages. 1971. DM 26,-

Vol. 244: Séminaire Bourbaki - vol. 1970/71. Exposés 382-399. IV, 356 pages. 1971. DM 26,-

Vol. 245: D. E. Cohen, Groups of Cohomological Dimension One. V, 99 pages. 1972. DM 16,-

Vol. 246: Lectures on Rings and Modules. Tulane University Ring and Operator Theory Year, 1970-1971. Volume I. X, 661 pages. 1972. DM 40,-

Vol. 247: Lectures on Operator Algebras. Tulane University Ring and Operator Theory Year, 1970-1971. Volume II. XI, 786 pages. 1972. DM 40,-

Vol. 248: Lectures on the Applications of Sheaves to Ring Theory. Tulane University Ring and Operator Theory Year, 1970-1971. Volume III. VIII, 315 pages. 1971. DM 26,-

Vol. 249: Symposium on Algebraic Topology. Edited by P. J. Hilton. VII, 111 pages. 1971. DM 16,-

Vol. 250: B. Jónsson, Topics in Universal Algebra. VI, 220 pages. 1972. DM 20,-

Vol. 251: The Theory of Arithmetic Functions. Edited by A. A. Gioia and D. L. Goldsmith VI, 287 pages. 1972. DM 24,-

Vol. 252: D. A. Stone, Stratified Polyhedra. IX, 193 pages. 1972. DM 18,-

Vol. 253: V. Komkov, Optimal Control Theory for the Damping of Vibrations of Simple Elastic Systems. V, 240 pages. 1972. DM 20,-

Vol. 254: C. U. Jensen, Les Foncteurs Dérivés de $\varprojlim$ et leurs Applications en Théorie des Modules. V, 103 pages. 1972. DM 16,-

Vol. 255: Conference in Mathematical Logic - London '70. Edited by W. Hodges. VIII, 351 pages. 1972. DM 26,-

Vol. 256: C. A. Berenstein and M. A. Dostal, Analytically Uniform Spaces and their Applications to Convolution Equations. VII, 130 pages. 1972. DM 16,-

Vol. 257: R. B. Holmes, A Course on Optimization and Best Approximation. VIII, 233 pages. 1972. DM 20,-

Vol. 258: Séminaire de Probabilités VI. Edited by P. A. Meyer. VI, 253 pages. 1972. DM 22,-

Vol. 259: N. Moulis, Structures de Fredholm sur les Variétés Hilbertiennes. V, 123 pages. 1972. DM 16,-

Vol. 260: R. Godement and H. Jacquet, Zeta Functions of Simple Algebras. IX, 188 pages. 1972. DM 18,-

Vol. 261: A. Guichardet, Symmetric Hilbert Spaces and Related Topics. V, 197 pages. 1972. DM 18,-

Vol. 262: H. G. Zimmer, Computational Problems, Methods, and Results in Algebraic Number Theory. V, 103 pages. 1972. DM 16,-

Vol. 263: T. Parthasarathy, Selection Theorems and their Applications. VII, 101 pages. 1972. DM 16,-

Vol. 264: W. Messing, The Crystals Associated to Barsotti-Tate Groups: With Applications to Abelian Schemes. III, 190 pages. 1972. DM 18,-

Vol. 265: N. Saavedra Rivano, Catégories Tannakiennes. II, 418 pages. 1972. DM 26,-

Vol. 266: Conference on Harmonic Analysis. Edited by D. Gulick and R. L. Lipsman. VI, 323 pages. 1972. DM 24,-

Vol. 267: Numerische Lösung nichtlinearer partieller Differential- und Integro-Differentialgleichungen. Herausgegeben von R. Ansorge und W. Törnig, VI, 339 Seiten. 1972. DM 26,-

Vol. 268: C. G. Simader, On Dirichlet's Boundary Value Problem. IV, 238 pages. 1972. DM 20,-

Vol. 269: Théorie des Topos et Cohomologie Etale des Schémas. (SGA 4). Dirigé par M. Artin, A. Grothendieck et J. L. Verdier. XIX, 525 pages. 1972. DM 50,-

Vol. 270: Théorie des Topos et Cohomologie Etale des Schémas. Tome 2. (SGA 4). Dirigé par M. Artin, A. Grothendieck et J. L. Verdier. V, 418 pages. 1972. DM 50,-

Vol. 271: J. P. May, The Geometry of Iterated Loop Spaces. IX, 175 pages. 1972. DM 18,-

Vol. 272: K. R. Parthasarathy and K. Schmidt, Positive Definite Kernels, Continuous Tensor Products, and Central Limit Theorems of Probability Theory. VI, 107 pages. 1972. DM 16,-

Vol. 273: U. Seip, Kompakt erzeugte Vektorräume und Analysis. IX, 119 Seiten. 1972. DM 16,-

Vol. 274: Toposes, Algebraic Geometry and Logic. Edited by. F. W. Lawvere. VI, 189 pages. 1972. DM 18,-

Vol. 275: Séminaire Pierre Lelong (Analyse) Année 1970-1971. VI, 181 pages. 1972. DM 18,-

Vol. 276: A. Borel, Représentations de Groupes Localement Compacts. V, 98 pages. 1972. DM 16,-

Vol. 277: Séminaire Banach. Edité par C. Houzel. VII, 229 pages. 1972. DM 20,-

Vol. 278: H. Jacquet, Automorphic Forms on GL(2). Part II. XIII, 142 pages. 1972. DM 16,–

Vol. 279: R. Bott, S. Gitler and I. M. James, Lectures on Algebraic and Differential Topology. V, 174 pages. 1972. DM 18,–

Vol. 280: Conference on the Theory of Ordinary and Partial Differential Equations. Edited by W. N. Everitt and B. D. Sleeman. XV, 367 pages. 1972. DM 26,–

Vol. 281: Coherence in Categories. Edited by S. Mac Lane. VII, 235 pages. 1972. DM 20,–

Vol. 282: W. Klingenberg und P. Flaschel, Riemannsche Hilbertmannigfaltigkeiten. Periodische Geodätische. VII, 211 Seiten. 1972. DM 20,–

Vol. 283: L. Illusie, Complexe Cotangent et Déformations II. VII, 304 pages. 1972. DM 24,–

Vol. 284: P. A. Meyer, Martingales and Stochastic Integrals I. VI, 89 pages. 1972. DM 16,–

Vol. 285: P. de la Harpe, Classical Banach-Lie Algebras and Banach-Lie Groups of Operators in Hilbert Space. III, 160 pages. 1972. DM 16,–

Vol. 286: S. Murakami, On Automorphisms of Siegel Domains. V, 95 pages. 1972. DM 16,–

Vol. 287: Hyperfunctions and Pseudo-Differential Equations. Edited by H. Komatsu. VII, 529 pages. 1973. DM 36,–

Vol. 288: Groupes de Monodromie en Géométrie Algébrique. (SGA 7 I). Dirigé par A. Grothendieck. IX, 523 pages. 1972. DM 50,–

Vol. 289: B. Fuglede, Finely Harmonic Functions. III, 188. 1972. DM 18,–

Vol. 290: D. B. Zagier, Equivariant Pontrjagin Classes and Applications to Orbit Spaces. IX, 130 pages. 1972. DM 16,–

Vol. 291: P. Orlik, Seifert Manifolds. VIII, 155 pages. 1972. DM 16,–

Vol. 292: W. D. Wallis, A. P. Street and J. S. Wallis, Combinatorics: Room Squares, Sum-Free Sets, Hadamard Matrices. V, 508 pages. 1972. DM 50,–

Vol. 293: R. A. DeVore, The Approximation of Continuous Functions by Positive Linear Operators. VIII, 289 pages. 1972. DM 24,–

Vol. 294: Stability of Stochastic Dynamical Systems. Edited by R. F. Curtain. IX, 332 pages. 1972. DM 26,–

Vol. 295: C. Dellacherie, Ensembles Analytiques, Capacités, Mesures de Hausdorff. XII, 123 pages. 1972. DM 16,–

Vol. 296: Probability and Information Theory II. Edited by M. Behara, K. Krickeberg and J. Wolfowitz. V, 223 pages. 1973. DM 20,–

Vol. 297: J. Garnett, Analytic Capacity and Measure. IV, 138 pages. 1972. DM 16,–

Vol. 298: Proceedings of the Second Conference on Compact Transformation Groups. Part 1. XIII, 453 pages. 1972. DM 32,–

Vol. 299: Proceedings of the Second Conference on Compact Transformation Groups. Part 2. XIV, 327 pages. 1972. DM 26,–

Vol. 300: P. Eymard, Moyennes Invariantes et Représentations Unitaires. II. 113 pages. 1972. DM 16,–

Vol. 301: F. Pittnauer, Vorlesungen über asymptotische Reihen. VI, 186 Seiten. 1972. DM 18,–

Vol. 302: M. Demazure, Lectures on p-Divisible Groups. V, 98 pages. 1972. DM 16,–

Vol. 303: Graph Theory and Applications. Edited by Y. Alavi, D. R. Lick and A. T. White. IX, 329 pages. 1972. DM 26,–

Vol. 304: A. K. Bousfield and D. M. Kan, Homotopy Limits, Completions and Localizations. V, 348 pages. 1972. DM 26,–

Vol. 305: Théorie des Topos et Cohomologie Etale des Schémas. Tome 3. (SGA 4). Dirigé par M. Artin, A. Grothendieck et J. L. Verdier. VI, 640 pages. 1973. DM 50,–

Vol. 306: H. Luckhardt, Extensional Gödel Functional Interpretation. VI, 161 pages. 1973. DM 18,–

Vol. 307: J. L. Bretagnolle, S. D. Chatterji et P.-A. Meyer, Ecole d'été de Probabilités: Processus Stochastiques. VI, 198 pages. 1973. DM 20,–

Vol. 308: D. Knutson, λ-Rings and the Representation Theory of the Symmetric Group. IV, 203 pages. 1973. DM 20,–

Vol. 309: D. H. Sattinger, Topics in Stability and Bifurcation Theory. VI, 190 pages. 1973. DM 18,–

Vol. 310: B. Iversen, Generic Local Structure of the Morphisms in Commutative Algebra. IV, 108 pages. 1973. DM 16,–

Vol. 311: Conference on Commutative Algebra. Edited by J. W. Brewer and E. A. Rutter. VII, 251 pages. 1973. DM 22,–

Vol. 312: Symposium on Ordinary Differential Equations. Edited by W. A. Harris, Jr. and Y. Sibuya. VIII, 204 pages. 1973. DM 22,–

Vol. 313: K. Jörgens and J. Weidmann, Spectral Properties of Hamiltonian Operators. III, 140 pages. 1973. DM 16,–

Vol. 314: M. Deuring, Lectures on the Theory of Algebraic Functions of One Variable. VI, 151 pages. 1973. DM 16,–

Vol. 315: K. Bichteler, Integration Theory (with Special Attention to Vector Measures). VI, 357 pages. 1973. DM 26,–

Vol. 316: Symposium on Non-Well-Posed Problems and Logarithmic Convexity. Edited by R. J. Knops. V, 176 pages. 1973. DM 18,–

Vol. 317: Séminaire Bourbaki – vol. 1971/72. Exposés 400–417. IV, 361 pages. 1973. DM 26,–

Vol. 318: Recent Advances in Topological Dynamics. Edited by A. Beck. VIII, 285 pages. 1973. DM 24,–

Vol. 319: Conference on Group Theory. Edited by R. W. Gatterdam and K. W. Weston. V, 188 pages. 1973. DM 18,–

Vol. 320: Modular Functions of One Variable I. Edited by W. Kuyk. V, 195 pages. 1973. DM 18,–

Vol. 321: Séminaire de Probabilités VII. Edité par P. A. Meyer. VI, 322 pages. 1973. DM 26,–

Vol. 322: Nonlinear Problems in the Physical Sciences and Biology. Edited by I. Stakgold, D. D. Joseph and D. H. Sattinger. VIII, 357 pages. 1973. DM 26,–

Vol. 323: J. L. Lions, Perturbations Singulières dans les Problèmes aux Limites et en Contrôle Optimal. XII, 645 pages. 1973. DM 42,–

Vol. 324: K. Kreith, Oscillation Theory. VI, 109 pages. 1973. DM 16,–

Vol. 325: Ch.-Ch. Chou, La Transformation de Fourier Complexe et L'Equation de Convolution. IX, 137 pages. 1973. DM 16,–

Vol. 326: A. Robert, Elliptic Curves. VIII, 264 pages. 1973. DM 22,–

Vol. 327: E. Matlis, 1-Dimensional Cohen-Macaulay Rings. XII, 157 pages. 1973. DM 18,–

Vol. 328: J. R. Büchi and D. Siefkes, The Monadic Second Order Theory of All Countable Ordinals. VI, 217 pages. 1973. DM 20,–

Vol. 329: W. Trebels, Multipliers for (C, α)-Bounded Fourier Expansions in Banach Spaces and Approximation Theory. VII, 103 pages. 1973. DM 16,–

Vol. 330: Proceedings of the Second Japan-USSR Symposium on Probability Theory. Edited by G. Maruyama and Yu. V. Prokhorov. VI, 550 pages. 1973. DM 36,–

Vol. 331: Summer School on Topological Vector Spaces. Edited by L. Waelbroeck. VI, 226 pages. 1973. DM 20,–

Vol. 332: Séminaire Pierre Lelong (Analyse) Année 1971-1972. V, 131 pages. 1973. DM 16,–

Vol. 333: Numerische, insbesondere approximationstheoretische Behandlung von Funktionalgleichungen. Herausgegeben von R. Ansorge und W. Törnig. VI, 296 Seiten. 1973. DM 24,–

Vol. 334: F. Schweiger, The Metrical Theory of Jacobi-Perron Algorithm. V, 111 pages. 1973. DM 16,–

Vol. 335: H. Huck, R. Roitzsch, U. Simon, W. Vortisch, R. Walden, B. Wegner und W. Wendland, Beweismethoden der Differentialgeometrie im Großen. IX, 159 Seiten. 1973. DM 18,–

Vol. 336: L'Analyse Harmonique dans le Domaine Complexe. Edité par E. J. Akutowicz. VIII, 169 pages. 1973. DM 18,–

Vol. 337: Cambridge Summer School in Mathematical Logic. Edited by A. R. D. Mathias and H. Rogers. IX, 660 pages. 1973. DM 42,–

Vol. 338: J. Lindenstrauss and L. Tzafriri, Classical Banach Spaces. IX, 243 pages. 1973. DM 22,–

Vol. 339: G. Kempf, F. Knudsen, D. Mumford and B. Saint-Donat, Toroidal Embeddings I. VIII, 209 pages. 1973. DM 20,–

Vol. 340: Groupes de Monodromie en Géométrie Algébrique. (SGA 7 II). Par P. Deligne et N. Katz. X, 438 pages. 1973. DM 40,–

Vol. 341: Algebraic K-Theory I, Higher K-Theories. Edited by H. Bass. XV, 335 pages. 1973. DM 26,–

Vol. 342: Algebraic K-Theory II, "Classical" Algebraic K-Theory, and Connections with Arithmetic. Edited by H. Bass. XV, 527 pages. 1973. DM 36,–